# FIELD GUIDE TO THE
# Birds of
# Trinidad & Tobago

Martyn Kenefick, Robin Restall
and Floyd Hayes

YALE UNIVERSITY PRESS
NEW HAVEN AND LONDON

*To Richard ffrench*
*For the legacy of his seminal book 'A Guide to the Birds of Trinidad & Tobago',*
*the various papers he has published about the birds of these islands,*
*his kindness and friendship, and his good wishes in this venture,*
*which we trust will complement his past, present and future contributions*
*to the ornithology of Trinidad & Tobago*

Published 2007 in the United Kingdom by Christopher Helm, an imprint of A & C Black Publishers Ltd.
Published 2008 in the United States by Yale University Press.

Commissioning editor: Nigel Redman
Production and design: Julie Dando, Fluke Art, Cornwall

Printed and bound by C&C Offset, China

Library of Congress Control Number: 2007936970
ISBN 978-0-300-13557-2 (pbk. : alk. paper)

A catalogue record for this book is available from the British Library.

The paper in this book meets the guidelines for permanence and durability of the Committee on Production Guidelines for Book Longevity of the Council on Library Resources.

10 9 8 7 6 5 4 3 2 1

# CONTENTS

# INTRODUCTION

Located on the continental shelf of north-eastern South America, the tropical islands of Trinidad and Tobago are host to an exceptionally rich and diverse avifauna. Their close proximity to South America brings additional species visiting from the continental mainland. Situated at the southern end of a migratory flyway, which follows a stepping-stone chain of islands between North America and South America, migrants from North America have also been recorded, whilst Eurasian and African vagrants have reached Trinidad & Tobago by crossing the Atlantic Ocean, assisted by either trade winds or ships. Finally, human activities have altered the native avifauna, causing the extirpation of several species of birds through habitat destruction and trapping, and introducing numerous exotic species.

Trinidad & Tobago have long been a popular birding destination for visiting birders who enjoy the sheer abundance of birds and relative ease of access to a diversity of habitats. Birding, the hobby of searching for and identifying birds, and ornithology, the scientific study of birds, are both rapidly growing in popularity. They are challenging pursuits based upon the accurate identification of birds. The chief purpose of this guide, therefore, is to provide a portable tool equipping birders and ornithologists alike with the information required to accurately identify birds in the field. Consequently, we have illustrated and described every species known to occur naturally or to have been successfully introduced to Trinidad & Tobago up to mid-2007, with Orchard Oriole being the most recent confirmed addition.

The vast majority of the illustrations have been taken from *Birds of Northern South America: An Identification Guide* (Restall *et al.* 2006), but several images have been repainted or corrected as necessary; in our ever-changing taxonomic world, it is important that the correct subspecies occurring on the islands are depicted. The text for the field guide is new, but because of inherent space constraints we have limited the text for each species to the most essential information to identify a bird. This does not imply that study of the natural history of each species is less important than identification, but rather that it is simply beyond the scope of this book. We fervently hope this field guide will inspire others, especially local residents, to carefully study and document the status and natural history of the birds of Trinidad & Tobago, thereby augmenting our scientific knowledge. Obtaining such information is vital for implementing conservation strategies to preserve the rich but increasingly imperiled birdlife.

# ACKNOWLEDGEMENTS

We acknowledge, with deep thanks and full appreciation, the help in various ways we received from the following people and institutions during the production of this book: Mark Berres, Stephen Bodnar, Kenny Calderon, Dave Cooper, Steven Easley, Jack Eitniear, Richard ffrench, Wordsworth Frank, Theodore Garnett, Newton George, Geoffrey Gomes, Brett Hayes, William Hayes, Cecilia Herrera, Adolphus James, Gladwyn James, Howard Kilpatrick, Guy Kirwan, the late Nedra Klein, Nigel Lallsingh, Keisha Lallsingh, Floyd Lucas, Charlie Madoo, the late James Madoo, Sean Madoo, Tony Marr, William Murphy, Roger Neckles, John O'Neill, Carol Ramjohn, Dave Ramlal, Roodal Ramlal, Jogie Ramlal, Mahese Ramlal, Nigel Redman, Courtenay Rooks, Ishmaelangelo Samad, Bryan Sanasie, Chris Sharpe, Neville Trimm, Graham White, the American Bird Conservancy, Amoco Trinidad Oil Company, Asa Wright Nature Centre, BirdLife International, British Petroleum, Caribbean Union College (now University of the Southern Caribbean), Center for the Study of Tropical Birds, Fauna & Flora International, Lincoln Park Zoo, Republic Bank Limited, Trinidad & Tobago National Petroleum Marketing Company Ltd, University of the West Indies, and Trinmar Ltd.

The plates were largely taken from *Birds of Northern South America*, and the task of re-composing the images to produce the plates for this book was again undertaken by Julie Dando. She also made many corrections on-screen and her skill, patience and professionalism in dealing with all the picky little comments from the authors is remarkable. She deserves special thanks and recognition.

A very special thank you is long overdue to our patient and long-suffering wives, Petra, Mariela and Marta, for their tolerance, support and understanding throughout.

# GEOGRAPHY

Trinidad and Tobago are continental islands that were formerly part of the South American continent. Indeed, a land bridge may have connected south-western Trinidad to the mainland as recently as 1,500 years ago, based on studies of fossil coral reefs off north-western Trinidad which could only have formed during more oceanic conditions than exist at present (Kenny 1995). Both islands were almost certainly connected to the continent during the last ice age, *c.*10,000–14,000 years ago, when sea levels were considerably lower than now (Comeau 1991).

## Trinidad

Trinidad is a relatively large 'flake' of South America, comprising 4,520 km². It is separated from the Venezuelan mainland by the Gulf of Paria. At the northern end of the gulf, the north-western or Chaguaramas Peninsula of Trinidad is separated from the Paria Peninsula of Venezuela by the notorious Boca de la Serpiente, or Dragon's Mouth Strait. The Dragon's Mouth is studded with a set of 'teeth' collectively referred to as the Bocas Islands, which comprises five major islands (Patos, Chacachacare, Huevos, Monos and Gaspar Grande) and many smaller islets. In the south of the gulf, the Columbus Channel separates the south-western or Icacos Peninsula of Trinidad from Venezuela by less than 20 km. Soldado Rock, a small islet, is situated off the tip of the Icacos Peninsula.

Trinidad possesses three distinct chains of mountains, each oriented west to east. The Northern Range is the highest, attaining a maximum height of 925 m, on Cerro del Aripo. The Central Range rises to just 308 m, at Mt Tamana, and the Southern Range reaches 304 m in the Trinity Hills. Extensive swamps occur at the mouths of rivers draining these uplands, including Caroni

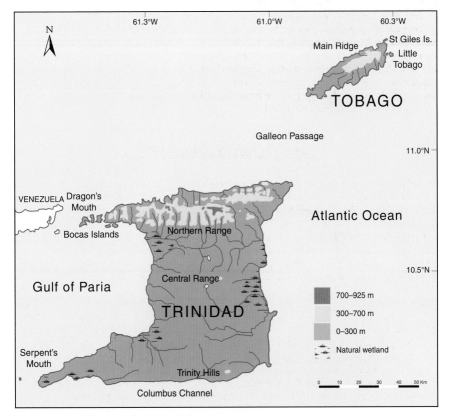

Major topograpical features of Trinidad and Tobago

Swamp, Oropouche Lagoon and, formerly, Los Blanquizales on the west and south coasts, and Nariva Swamp on the east. High sediment loads along Trinidad's coasts limit coral reefs to small, scattered patches off the north coast.

## Tobago

Tobago is a 306 km² island separated from Trinidad, to the south-west, by the 36 km-wide Galleon Passage. Several small islets, including St Giles Island, Goat Island and Little Tobago, are situated off Tobago's north-east tip. Tobago is geologically related to the southern Caribbean Islands of Aruba, Bonaire and Curaçao, but has been displaced eastward.

A single mountain range, the Main Ridge, forms the backbone of Tobago and attains a maximum height of 576 m on Centre Hill. The relatively flat south-west of Tobago comprises an ancient coralline platform uplifted above sea level. The only extensive swamps occur near the south-western tip of the island; remnants of Buccoo Swamp occur on the north side and Kilgwyn Swamp to the south. Coral reefs line the coasts of Tobago and are most extensive at Buccoo Reef in the south-west of the island.

# CLIMATE

Trinidad and Tobago are located *c*.10–11°N of the equator, well within the tropics (Berridge 1981). Typical of tropical islands, temperature variation is relatively minor due to intense solar radiation throughout the year, combined with the buffering effects of high humidity and their oceanic setting. Temperatures are coolest during the northern winter, especially on clear nights in the dry season when humidity is lowest, and warmest in late spring, especially during sunny days late in the dry season. Mean temperatures are lowest during January (24.5°C/76°F at Piarco, Trinidad) and highest in May (26.6°C/80°F). Temperatures average slightly cooler during the northern summer because of frequent cloud cover and rainfall, and the buffering effects of high humidity during the wet season. (Mean temperatures from Piarco give a totally false impression, as the thermometer is way up in the air, and probably in the shade! Almost invariably the temperature at ground level is 32–34°C (89–93°F) and the heat index 35–37°C (95–98°F). This is relevant information to a visiting birder.

In the tropics, moist air masses are warmed by direct sunshine and rise high into the atmosphere, where they cool and expand outward towards the poles. As the air masses cool, moisture condenses and gives way to precipitation. The inter-tropical convergence is formed by the conjunction of rising air masses, which shift north and south as the sun moves between the Tropic of Cancer in June and the Tropic of Capricorn in December. Due to these shifts, rainfall variation is much more pronounced than temperature, with two distinct seasons. The dry season typically commences in January and terminates in late May, with March being the driest month (mean 3.4 cm at Piarco). Many bush fires occur in this period, virtually all set by humans. Although more than a month may

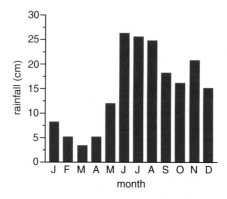

Mean monthly rainfall on Trinidad, based on Berridge (1981).

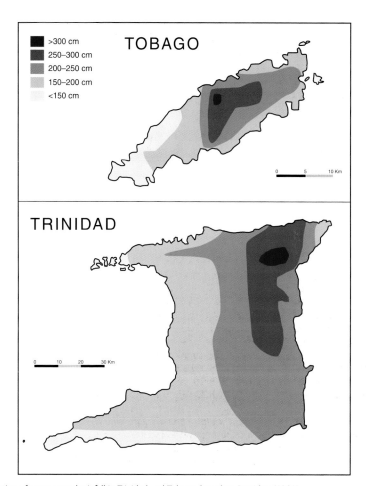

Legend:
- >300 cm
- 250–300 cm
- 200–250 cm
- 150–200 cm
- <150 cm

TOBAGO

0    5    10 Km

TRINIDAD

0    10    20    30 Km

Distribution of mean annual rainfall in Trinidad and Tobago, based on Berridge (1981).

pass without rainfall, humidity remains relatively high year-round. The wet season typically begins with heavy showers in late May and extends until December, with June being the wettest month (mean 25.4 cm). A short break in the wet season, essentially a mini dry season, locally referred to as the *petite carime*, typically occurs in September or October. Mention should be made of mean humidity, which is over 50% in the dry season and over 90% during the rains.

Strong easterly trade winds predominate for most of the year, but winds often blow from the west during winter. Precipitation is relatively low along the east coasts of both islands, but as the predominant easterly trade winds rise and cool during their passage over the highest elevations, rainfall steadily increases with altitude. Mean annual rainfall may exceed 250 cm at highest elevations on Trinidad. The mountains wring out humidity from the air and form a rain shadow to the west, where precipitation rapidly declines as drier air descends and flows towards the Caribbean. Chacachacare, the westernmost of the Bocas Islands, averages only 115 cm of rainfall per year.

Cyclonic storm systems, which are nature's way of removing heat from the tropics, typically form in the eastern tropical Atlantic Ocean during the hurricane season of June–November. As they strengthen whilst hurtling west towards Trinidad & Tobago, they almost invariably veer north toward the Lesser Antilles, and only rarely strike Trinidad &Tobago. The last major hurricane to hit the islands was Hurricane Flora, which scored a direct hit on Tobago in 1963, though Hurricane Ivan caused considerable damage and loss of life on Tobago as recently as 2004.

# HABITATS

The natural plant communities of a given area are generally determined by three abiotic factors: temperature, humidity and soil. Because temperature varies little in Trinidad & Tobago, humidity and, to a lesser extent, soil are the more important factors determining vegetation. In contrast to plants, the animal communities inhabiting a given area are more dependent on the species of plants and the vegetational structure, rather than abiotic factors. Vegetation communities tend to grade into one another and are thus difficult to classify. Although a wide variety of plant communities in Trinidad & Tobago have been described (Beard 1944, 1946), we have simplified the descriptions of these natural habitats.

At the highest elevations, where precipitation is greatest, **tropical rainforest** is the dominant habitat. Humidity is high year-round and adequate during the dry season to sustain the growth of plants. Many species of dicotyledonous trees flourish, with no species dominating, except along the highest ridges where a few species may do so. Rainforest trees grow very tall, with some emergents exceeding 50 m in height. Palms are common on Tobago but less so on Trinidad. The trunks of rainforest trees are usually straight, slender and unbranched up to 20 m above the forest floor. The canopy (upper layer of foliage) is dense and closed, permitting little light to penetrate to the forest floor. Smaller trees occur, providing several layers of canopies. Leaves are usually large and broad with entire margins and pointed 'drip' tips. Epiphytes – plants which grow on other plants – are ubiquitous on the limbs of trees and include various species of vines (lianas), bromeliads, aroids, orchids, ferns, cacti and mosses. Because nutrients are scarce in rainforest soils, root systems tend to be shallow, spreading laterally. Tree trunks often have wide buttresses, providing additional support. The tropical rainforests of Trinidad are often subdivided into various formations. **Cloud forest** or **elfin woodland** occurs on the highest peaks and ridges of the Northern Range, where fewer species of trees occur and those that do tend to be shorter and more twisted. **Montane rainforest** occurs on the highest slopes of the Northern Range whilst **lower montane rainforest** is found at intermediate to high elevations of all three ranges on Trinidad and the Main Ridge of Tobago.

Where less precipitation occurs at intermediate and low elevations of Trinidad & Tobago, **evergreen forest** or **seasonal forest** is the dominant vegetation. Rainfall is more seasonal than in tropical rainforest and humidity during the dry season is insufficient to maintain growth. Evergreen forest may closely resemble tropical rainforest, especially at higher elevations, but the trees tend to be shorter, and tree trunks tend to branch nearer the ground and to be less buttressed. Because more light often reaches the forest floor, there is usually more understorey vegetation. Epiphytes remain ubiquitous but are less abundant than in tropical rainforest. Leaves are mostly evergreen, persisting year-round.

At lower elevations in the rain shadow of western Trinidad and southern Tobago, and on the smaller offshore islets, **tropical dry forest** occurs where humidity is lowest. Trees are shorter than those of evergreen forest, epiphytes are less common, and because the canopy tends to be more open, understorey growth is often profuse. The leaves of most trees are deciduous and are shed during the dry season. Tropical dry forest is often subdivided into **semi-evergreen forest** at intermediate and low elevations, and **deciduous forest** at the lowest elevations and on offshore islets (where humidity is lowest).

In a few lowland areas of Trinidad, **tropical savanna** occurs, where a few species of grasses and sedges are the dominant plants on relatively flat terrain. Although sufficient rainfall occurs to support the growth of trees, the shallow soils of savannas are of relatively impervious clay or iron pan with poor drainage, thus overriding climatic factors by precluding the establishment of tree species. The ground is bumpy with numerous hummocks. Savannas tend to be waterlogged during the wet season and dry at other times. Because soil nutrients are scarce, carnivorous plants such as sundews *Drosera* spp. and bladderworts *Utricularia* spp. are common. The savannas form a mosaic with palm stands and forest, which occur on slightly higher terrain. In many areas of both main islands, savanna-like vegetation develops where the original forest has been removed, such as on the southern slopes of Trinidad's Northern Range.

Where the vegetation is exposed to strong winds and salty ocean spray, such as on beaches, sandbanks and cliffs along the coasts, **littoral woodland** is the predominant habitat. Trees are relatively small, reaching heights of up to 10 m, with distorted crowns and gnarled branches that often point away from the wind. Their leaves, which are adapted to the frequent salt spray, tend to be thick and fleshy with a waxy cuticle. The dominant species are sea grape *Coccoloba uvifera* and manchineel or poison apple *Hippomane mancinella*. This habitat is often used for nesting by colonial seabirds on small islands.

A variety of wetlands occur in Trinidad & Tobago. **Mangrove swamp** or **manglar** often occupies the intertidal zone at the mouths of streams and rivers along sheltered coasts. Although several species of trees are adapted to the unique conditions of these swamps, red mangrove *Rhizophora mangle* is the dominant species. It grows up to 25 m tall and is characterised by a network of stilt roots anchoring the tree in the mud. The leaves of mangrove trees are tough and succulent with entire margins and a thick, waxy cuticle. The most extensive mangrove swamps occur at Caroni on the west coast and Nariva on the east coast of Trinidad, and at Buccoo in western Tobago.

Freshwater or brackish **marshes** occur in several places on the coasts of Trinidad & Tobago, usually inland of mangrove swamps. They are characterised by sedges or grasses in relatively shallow water. In some areas, water lilies *Nymphaea ampla* and water hyacinths *Eichornia crassipes* are common. Some marshes are fairly extensive, especially at Caroni and Nariva on Trinidad, and at Buccoo on Tobago.

A variety of other wetland habitats are important for birds. **Beaches**, **mudflats** and **sandbars** form the margins of many coastal areas and freshwater wetlands. Beaches are most extensive along the east coast of Trinidad, whilst mudflats are most extensive on the west-central coast of Trinidad. **Open water** is provided by streams, rivers, ponds, reservoirs, lagoons, bays and the ocean. Some of the smaller waterbodies dry up during the dry season.

Most of these natural habitats have been significantly degraded by human activities. However, some novel habitats have been formed entirely of exotic plants, including monoculture plantations of cacao *Theobroma cacao*, coffee *Swietenia macrophylla*, teak *Tectona grandis*, Caribbean pine *Pinus caribbaea*, sugarcane *Saccharum* spp. and rice *Oryza sativa*. Biodiversity is generally low in these, except in rice plantations, which attract more birds than natural marshes, and in cacao and coffee plantations where there is an overstorey of native or exotic trees.

# TAXONOMY AND NOMENCLATURE

Based on shared morphological traits representing evolutionary relationships, species are grouped into a hierarchical classification system. At the highest levels, birds belong to the kingdom Animalia (all animals), phylum Chordata (animals with a notochord or backbone) and class Aves (all birds), respectively. At lower levels, birds are grouped into orders, families and genera, respectively. The order Passeriformes, for example, includes many families of perching birds such as flycatchers (family Tyrannidae). A family comprises one or more genera and a genus includes one or more species.

## Species and Subspecies

The definition of a species has vexed scientists for centuries. Currently most ornithologists embrace the Biological Species Concept, which defines a species as a group of freely interbreeding natural populations that is essentially (not necessarily completely) reproductively isolated from other such groups. The operational term here is 'freely interbreeding', which implies 'reproductive isolation' from other species. But because populations of birds are often in the process of diverging and processes are difficult to evaluate, it is often difficult to determine whether two slightly different populations belong to one or two species. Hybridisation between bird species is widespread, even between species that are not each other's closest relatives, and its occurrence between two species indicates that reproductive isolation is incomplete. If individuals of two distinct forms mate randomly with either form, they are freely interbreeding and are usually deemed to belong to a single species. But if individuals prefer to mate with their own form even though some mate with the other form, the two forms are not freely interbreeding and have attained essential reproductive isolation, indicating that they belong to different species.

The criterion of reproductive isolation is easiest to assess when two distinct forms co-exist. For example, the Green Heron *Butorides virescens* of North and Middle America and the Caribbean is similar to the Striated Heron *B. striata* of South America. Visual intermediates often occur where their ranges meet (e.g. on Tobago), suggesting that hybridisation occurs. But because intermediates comprise a small proportion of the population where the two forms meet, they probably prefer to mate with their own form, suggesting that they have achieved essential rather than complete reproductive isolation and have evolved into two distinct species (Hayes 2006). Another example is the Sandwich Tern *Sterna sandvicensis*. North American and northern Caribbean populations have a mostly black bill (*S. s. acuflavida*), whereas southern Caribbean and South American populations, known as Cayenne Tern (*S. [s.] eurygnatha*), have a largely yellow bill. Because interbreeding between dark-billed and yellow-billed individuals appears random in the few colonies where their ranges meet, and the transition from dark-billed to yellow-billed forms occurs gradually from north to south, the two forms probably lack reproductive isolation and may not have yet reached the level of species (Hayes 2004). For convenience, we treat them separately in this book.

The criterion of reproductive isolation is considerably more difficult to assess in cases where two distinct forms are geographically separated. Such cases necessitate that the degree of reproductive isolation be inferred by the similarity between the forms, based on morphology, vocalisations or genetics, and compared with differences between other closely related species. For example, the Trinidad Piping-Guan *Pipile pipile* is found only on Trinidad and is similar to populations in continental South America. However, several unique traits distinguish the Trinidad birds from those on the continent, and currently the Trinidad form is widely regarded as sufficiently different to warrant recognition as a species. The recognition of species by scientists is an imprecise science that often changes as new information concerning relationships is acquired.

Species are often subdivided into subspecies, which are geographically distinct races (the terms 'subspecies' and 'race' are often used interchangeably). But because variation amongst races is often trivial and may occur gradually (or clinally) along environmental gradients within a species' range, subdividing populations into well-defined subspecies is often difficult. Some currently recognised subspecies are distinctive and easily recognised; others appear so similar that individuals cannot be conclusively identified to subspecies. Although many subspecies have been formally named by ornithologists, some differ on mean morphological attributes (such as body size or shape) that may have an environmental rather than genetic basis, in which case they should not be recognised as subspecies. Extensive research may be required to understand the causes of geographic variation within a species, but even then it might remain a mystery.

## Bird Names

Scientists assign each species a scientific name comprising two words: the genus and the specific epithet. Scientific names are usually Latinised (but occasionally derived from Greek and other languages), and are always italicised, with the genus name capitalised and the specific epithet not so. In the case of subspecies, a third subspecific epithet is appended. For example, *Thamnophilus doliatus tobagensis*, the subspecies of Barred Antshrike occurring only on Tobago, is whiter below in the male and darker in the female than the race inhabiting Trinidad, *Thamnophilus doliatus fraterculus* (ffrench 1991).

Each species has only one scientific name that is unique and internationally recognised. However, scientific names sometimes change when scientists acquire new information about the relationships among birds. For example, two or more currently recognised species may be lumped into one species, or a single species may be split into two or more species. Furthermore, a species or suite of species may be removed from one genus to another when relationships between them become better understood.

Vernacular names, which are easier to remember than scientific names, are also applied to birds but these often vary, to a greater or lesser extent, from one place to another, and more than one name is often applied to a given species. For example, *Psarocolius decumanus* may be referred to locally as a 'cornbird', 'yellowtail' or 'pogga'. Several contending references have compiled 'official' English names of birds, which by convention are always treated as proper nouns by birders and ornithologists. For example, the official English name of *Psarocolius decumanus* is 'Crested Oropendola'. Treating English names of birds as proper nouns is used to distinguish a specific species from a non-specific species or group of species. For example, a Brown Booby is used to distinguish *Sula leucogaster* from a 'brown booby', which could refer to a brown immature of any of the three species of boobies occurring in Trinidad & Tobago.

In this guide, we have adopted the official English and scientific names used by the American Ornithologists' Union's (AOU) South American Checklist Committee, using British spellings. We have generally followed the AOU sequence, which is based upon evolutionary relationships, but have often departed from it, in order to illustrate species of similar appearance on the same page.

## REFERENCES

Beard, J. S. (1944) The natural vegetation of the island of Tobago, British West Indies. *Ecol. Monogr.* 14: 135–163.

Beard, J. S. (1946) *The Natural Vegetation of Trinidad.* Clarendon Press, Oxford.

Berridge, C. E. (1981) Climate. Pp. 2–12 *in* Cooper, G. C. & Bacon, P. R. (eds.) *The Natural Resources of Trinidad and Tobago.* Edward Arnold, London.

Comeau, P. L. (1991) Geological events influencing natural vegetation in Trinidad. *Living World, J. Trinidad & Tobago Field Nat. Cl.* 1991–92: 29–38.

ffrench, R. (1991) *A Guide to the Birds of Trinidad and Tobago.* Second edn. Cornell University Press, Ithaca, NY.

Hayes, F. E. (2004) Variability and interbreeding of Sandwich Terns and Cayenne Terns in the Virgin Islands, with comments on their systematic relationship. *N. Amer. Birds* 57: 566–572.

Hayes, F. E. (2006). Variation and hybridization in the Green Heron (*Butorides virescens*) and Striated Heron (*B. striata*) in Trinidad and Tobago, with comments on species limits. *J. Carib. Orn.* 19: 12–20.

Kenny, J. S. (1995) *Views from the bridge: a memoir of the freshwater fishes of Trinidad.* Trinprint Ltd, Barataria, Trinidad & Tobago.

Restall, R., Rodner, C. & Lentino, M. (2006) *Birds of Northern South America: An Identification Guide.* Christopher Helm, London

# BIRD IDENTIFICATION

## Learning to Identify Birds

Developing identification skills requires considerable study of birds both in the field and in a field guide. To become proficient, begin with familiar birds that are easily identified and search for predictable patterns that help you to quickly distinguish them from others. Then search for unfamiliar birds and practice studying and memorising the details of each bird, before searching for it in a field guide. Study the field guide to learn the terms used for the various body parts. Initially a bird may be quickly identified to a particular family (or even species) by a general impression of size and shape – often referred to as 'jizz' by birders.

Identifying or confirming the identity of a species begins by searching for specific clues to its identity, referred to as field marks. These may include relative size (which is often misjudged), shape and proportions, colours (of both feathers and the unfeathered parts such as the bill, orbital ring, eyes and legs), patterns, vocalisations and behaviour. In some species field marks are obvious and diagnostic; in others they may be subtle and difficult to perceive except under ideal viewing conditions, or may require observing from a particular angle or when the bird is flying. Search for multiple field marks which may be unique to a species, thus eliminating other possibilities.

Rather than relying on memory, serious birders often take field notes on unfamiliar birds by writing a description or quickly sketching the bird in question. Taking field notes forces you to study a bird more carefully and is more objective and reliable than memory alone when identifying a bird or documenting a rarity. Learning to take good field notes is an important step in becoming an expert birder.

Many birds, especially nocturnal species and those living in forests, are most quickly recognised and located by their vocalisations. Whilst some are best identified by their vocalisations, there are a number of species that can be recognised *only* by using voice. The most distinctive and complex vocalisations of birds are called songs. These are generally used to establish a territory and attract a mate. Shorter and simpler vocalisations are known as calls, and may function in a variety of behavioural contexts. For example, contact-calls facilitate communication between individuals and alarm-calls may communicate danger to others or startle a predator. Learning to identify birds by their vocalisations is usually more challenging than by their appearance; it is easier for some birders than for others. Vocalisations may be learned by listening to recordings of known birds or by hearing a song, then locating and identifying the source. Written descriptions of vocalisations can also be helpful.

## Individual Variation

Individuals of a given species differ slightly (sometimes dramatically) from each other in size, shape, coloration, voice and behaviour. Because of this, not all individuals will appear exactly like the illustrations in this guide, which portray the typical appearance of a bird or commonly observed variations. An understanding of the underlying causes of individual variation is essential for developing bird identification skills.

**Sex** Nearly all bird species are sexually dimorphic, which means the two sexes are morphologically different. In most species the sexes differ in average body size, with males in most – but certainly not all – families usually larger than females. But often the differences are slight with considerable overlap. In sexually monochromatic species coloration appears identical between the sexes. In some sexually dichromatic species the sexes differ subtly or even dramatically in colour, but often only during the breeding season. In most sexually dichromatic species males are more brightly coloured, usually due to sexual selection in which females prefer to mate with the brightest males. In some families, such as hummingbirds and manakins, older females may acquire male-like plumage. Several aspects of behavior may also differ between sexes. To give a few examples, males of most species sing more frequently (or even exclusively) than females, and the sexes may differ in their food preferences and in the timing of migration.

**Age and seasonal variation** The appearance of a bird always changes with age, and age-related changes are often linked to seasons. Recently hatched fledglings can change dramatically as they mature, first into juvenile, then immature and finally into adult plumage. Each species undergoes a sequence of moults and plumages (see Moults and Plumages below), often with different-coloured soft parts, before they become adult. Some attain adult appearance within a year of hatching, whereas others require two or more years. For example, large white-headed gulls usually require four or more years to attain adult plumage and soft-part coloration; because they moult partially twice per year, they acquire at least nine distinct plumages prior to becoming adult. The plumage of moulting birds often appears different from those that have completed moult. Some moulting birds have shorter wings or tails than normal, which could lead to confusion with other species. In some species adults may change subtly in colour as they become older.

**Polymorphism** Some species, e.g. Red-footed Booby *Sula sula* and Long-winged Harrier *Circus buffoni*, occur in more than one plumage colour – referred to as a colour morph or colour phase – that is unrelated to sex or age. The proportion of individuals belonging to each colour morph often varies in different parts of the range. This is analogous to the different eye colours of humans.

**Geographic variation** The body size, appearance and vocalisations of a species usually are not uniform throughout its range. Within a species, populations occurring in colder climates tend to be larger than those in warmer areas, even on a micro-geographic scale. For example, Bananaquit *Coereba flaveola* averages greater body mass at higher elevations in Trinidad & Tobago than those at lower elevations. Populations occurring in drier climates tend to be paler than those in more humid environments. Gradual (clinal) changes often occur along environmental gradients within contiguous populations on continents. Isolated (or disjunct) populations, such as those on islands, often differ slightly from other populations. In certain species, the populations on Tobago average larger body size or differ slightly in coloration from those on Trinidad. Some of these are sufficiently distinctive to warrant recognition as subspecies.

**Wear and fading** Bird feathers may become tatty with wear and bleached by the sun, and can appear markedly different from freshly moulted feathers. Worn feathers appear frayed and paler than normal ones. Pale feathers lacking melanin, which strengthens feathers, tend to wear more quickly than darker feathers. Species living in open environments where they are exposed to the sun tend to fade more rapidly than those in forests. The feathers of juveniles tend to wear and fade more rapidly than those of adults, and the same is true of the primary feathers on the wing.

**Adventitious colouring** The bill, legs and feathers of birds may be stained by elements in the environment. For example, waterbirds, especially the paler species such as egrets, often acquire mud-staining on the bill, legs, and feathers of the head and underparts. Rust-staining from iron compounds, such as the orange stains on the faces of some migrant shorebirds, also occurs. Some waterbirds become stained by oil spills, which occur frequently around Trinidad & Tobago. The heads of frugivorous or nectarivorous birds are sometimes stained by pollen, and the underparts of incubating birds may be stained by nesting material.

**Aberrant coloration** The appearance of a small proportion of birds is abnormal due to environmental or genetic causes, or even an interaction between the two, and occurs more frequently in some species than others. For example, Yellow-legged Thrushes *Platycichla flavipes* in Tobago have a relatively high frequency of white feathering, especially on the head.

Albinism is a rare hereditary condition (caused by homozygous recessive alleles) resulting in the complete absence of pigments in the plumage and soft parts. Albinos are recognised by a combination of white plumage and pink bill, eyes and legs.

Leucism, the loss of pigments in feathers but not soft parts, is the most common form of aberrant coloration, resulting in white or creamy feathers. It may be restricted to as few as a single feather (partial leucism, often incorrectly called partial albinism) or include all feathers (complete leucism). The whitish bases of normal feathers are sometimes exposed during moult and may be mistaken for leucism. Leucism is often confused with dilution, which refers to the even reduction in quantity of all pigments present on a given bird.

Melanism, the excessive deposition of darker eumelanin pigments in feathers, occurs less frequently than either albinism or leucism, and results in an unusually dark individual. Melanism also may be complete or incomplete. In some species, such as Long-winged Harrier and Short-tailed Hawk *Buteo brachyurus*, it occurs sufficiently frequently to be considered normal, and such birds are termed 'dark morphs'.

Carotenism refers to any abnormality of intensity or distribution of carotenoid pigments. It is usually caused by schizochroism, which is the absence of one pigment that exposes the underlying pigments. For example, pure yellow Orange-winged Parrots *Amazona amazonica* are occasionally seen in Trinidad & Tobago; because these birds lack melanin pigments, the underlying carotenoid pigments are exposed, resulting in an all-yellow appearance.

**Bill and feet deformities** Birds occasionally suffer from an injury, genetic defect or disease resulting in deformity of the bill or feet. Birds' bills grow continuously but are normally worn down by use, preventing it from growing larger. Injuries to the tip of the bill or, more rarely, a genetic defect may cause the bill to grow unusually long or twisted. Such birds may alter their behaviour in order to feed. Several diseases may cause growths to occur on the feet of birds, especially amongst those species that forage on the ground. Legs, claws or toes are occasionally broken, distorted or naturally amputated, often due to encounters with man-made objects.

**Hybrids and intergrades** Some define hybridisation as the interbreeding of two species, in which case the offspring are called hybrids, and refer to the interbreeding of two subspecies as intergradation, in which case the offspring are called intergrades. Others define hybridisation as the interbreeding of morphologically distinct populations in secondary contact, regardless of taxonomic rank (such as between subspecies). Neither definition is wholly satisfactory. The first is difficult to apply when taxonomic rank is uncertain and the genetic processes occurring between interbreeding species is the same as that occurring between interbreeding subspecies. The second can be taken to the extreme to apply to populations with trivial morphological differences, such as interbreeding between larger individuals at higher elevations and smaller individuals at lower elevations.

Regardless of definition, being intermediate between the two parental taxa, exhibiting traits of both forms, best recognises hybrids. Hybrids are also more variable than either parental taxon. Furthermore, hybrids often backcross with either parental taxon, resulting in offspring that closely resemble one parent and forming a continuum of variation between the two taxa. Hybrids are generally rare in nature, but occur more frequently in some groups than others (e.g. ducks, gulls, hummingbirds). In Trinidad & Tobago, presumed hybrids between Green and Striated Herons, and between the northern Sandwich Tern and southern Cayenne Tern are often seen.

Sometimes it is difficult to assess whether intermediate individuals are hybrids or extreme variants of one species. For example, some tanagers in Trinidad & Tobago appear intermediate between Palm Tanager *Thraupis palmarum* and Blue-grey Tanager *T. episcopus*, and have been presumed to be hybrids. However, careful examination of large series of specimens has revealed these birds to be aberrantly plumaged Palm Tanagers.

## Glossary

Other terms used in the book are as follows.

**Austral** Pertaining to or coming from the south (i.e. South America).
**Arched** Wings curving upwards to the carpal, with flight feathers drooping down.
**Bowed** Wings curving or drooping gently below the horizontal.
**Carpal** The bend of the wing.
**Carpal bar** A contrasting diagonal stripe from the wrist of the wing inward towards the body.
**Cere** A bare patch of often brightly coloured skin between the bill and lores in some birds (e.g. raptors).
**Chevrons** V-shaped markings, usually along the flanks.
**Coronal bands** Stripes on the side of the crown (also called lateral crown stripes).
**Crepuscular** Active at dawn and dusk.
**Cupped** Curved wings; inner wing raised and outer wing drooped.
**Dihedral** Wings held in a raised V shape, quite frequent in some raptors.
**Dimorphic** Two colour morphs, e.g. in several herons.
**Diurnal** Active by day, relevant to some owls which may be either partially or exclusively diurnal.
**Eclipse** Post-breeding moult plumage of (usually) male ducks, which is female-like.
**Extirpated** (Locally) extinct.
**Feral** Free-flying and living in the wild, but originally of domestic stock.

**Gape** Strictly the mouth, but usually refers to bare skin at the corner of the mouth, most noticeable on juvenile birds.

**Gorget** Necklace of markings on the throat or upper breast.

**Gular sac** An inflatable patch of throat skin, as in male frigatebirds.

**Landing lights** Pale markings on the leading edge of the upperwing nearest the body.

**Mirrors** White subterminal spots on the outermost flight feathers of gulls.

**Morph** A distinct colour variation.

**Neotropics** The region encompassed by the islands of the Caribbean, Central and South America, from approximately central Mexico to Cape Horn.

**New World** North, Central and South America, together with islands of the Caribbean.

**Pectoral band** A band or clear colour demarcation across the breast.

**Polymorphic** More than two colour morphs.

**Postocular** Behind the eye, as in the white spot shown by many hummingbirds.

**Raptor** Any bird of prey.

**Scalloped** Pattern of semi-circular markings.

**Semipalmations** Webbing between the toes, found in many waders.

**Spangled** Covered with small spots (often iridescent, as in hummingbirds, giving a sparkling or glittering effect).

**Speculum** A (usually) square shape of prominent colour on the upper secondaries that is present in many ducks and parrots.

**Supraloral** A stripe or large spot across the top of the lores.

**Tertial crescent** Pale crescents on the folded wing, formed by pale tips on the tertials.

**Vermiculations** Very fine, often wavy, bars.

**Vinaceous/Vinous** A deep reddish, wine-like colour.

## Moults and Plumages

Plumage refers to the feathers of birds. Because feathers become frayed with age, they must be replaced periodically. Moult refers to the complicated process by which old feathers are replaced by new ones. A different plumage is formed whenever the freshly moulted feathers differ in appearance from those that have been replaced. The timing of moult has evolved to fit into an annual cycle that varies greatly amongst species and even among individuals of a species. It may occur quickly, within a matter of a few weeks, or be drawn out over several months. Because moult is energetically expensive, it usually occurs following the reproductive and migratory seasons, but may also be suspended during these periods. In the Neotropics, nearly all birds moult once annually and some undergo a complete moult annually, with a second partial moult usually prior to the breeding season. A few birds undergo three annual moults whilst, in contrast, some seabirds may take years to complete a single moult of the wings.

## Bird Topography

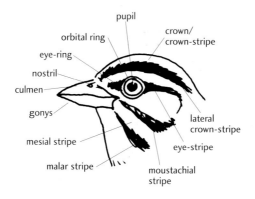

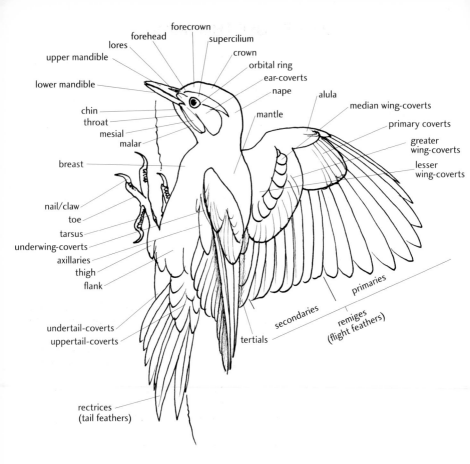

forecrown

forehead

supercilium

lores

crown

upper mandible

orbital ring

ear-coverts

lower mandible

nape

alula

median wing-coverts

chin

mantle

primary coverts

throat

mesial

greater
wing-coverts

malar

lesser
wing-coverts

breast

nail/claw

toe

tarsus

underwing-coverts

axillaries

thigh

flank

undertail-coverts

uppertail-coverts

primaries

secondaries

remiges
(flight feathers)

tertials

rectrices
(tail feathers)

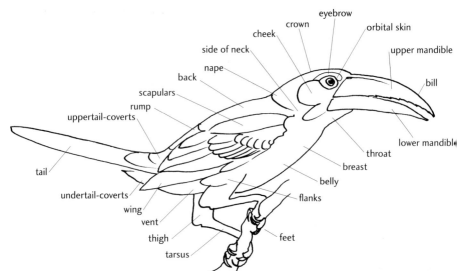

eyebrow

crown

orbital skin

cheek

side of neck

upper mandible

nape

back

bill

scapulars

rump

uppertail-coverts

lower mandible

throat

breast

tail

belly

undertail-coverts

flanks

wing

vent

thigh

feet

tarsus

18

# HOW TO USE THIS BOOK

Each family has a short introductory paragraph detailing some general characteristics common to each member of that family.

Each species account begins with the English and scientific names of the species. Monotypic species (birds which have no subspecies) are given a binomial scientific name (two words). Polytypic species are listed with three scientific names, the third name relating to the appropriate subspecies to be found in Trinidad & Tobago (if known). Where more than one subspecies occurs on the islands, the name of the more common race is listed and subsequently described. Other subspecies are detailed in the text.

## Measurements

The length of each species is given in centimetres. The wingspan (WS) is also given for a few groups such as seabirds and raptors.

## Species Account

The bulk of each account comprises a concise description of the species. Separate sections cover **Voice**, Similar Species (**SS**) and Status & Abundance (**Status**).

## Status

The general status of a species refers to its occurrence in its breeding and non-breeding ranges, which may change historically as populations colonise new areas or disappear from others. The following terms refer to the status of each species.

**Resident** A species known to breed within the region, based on observations of an active nest (with eggs or nestlings) or recently fledged young. For some species, breeding is suspected but unconfirmed.

**Migrant** A species that occurs during regular seasonal passage from its breeding range to its non-breeding range, usually from another continent such as North or South America. Such species may occur only on passage or throughout the non-breeding season, and more occasionally even year-round.

**Visitor** A species not known to breed but which may visit from nearby breeding areas or its normal range, such as adjacent South America or the Caribbean islands.

**Vagrant** A species that strays far from its normal range or migration route, e.g. between Europe and Africa.

**Extirpated** A formerly resident species which is now locally extinct.

## Abundance

The relative abundance of a species refers to its frequency of detection. Abundance varies with habitat. Any given species is more abundant in certain habitats and less abundant or absent in others. Species regarded as highly **local** occur only in specialised, geographically limited habitats, such as savannas. The abundance of many species also varies seasonally. Some may be present only at a particular time of year, e.g. spring migration; others may be present year-round, but are more abundant during a given season.

The following terms for relative abundance refer to the average frequency of detection (either seen or heard) within appropriate habitat at an appropriate season by an experienced observer.

**Abundant** 50 or more recorded daily.

**Common** Usually recorded daily.

**Uncommon** Usually recorded every 2–10 days.

**Scarce** Recorded less frequently than every ten days but more than five times per year.

**Rare** Not recorded more than five times per year.

**Very rare** Recorded every 1–10 years.

**Accidental** Recorded less than once per decade.

# DOCUMENTING AND REPORTING BIRD SIGHTINGS

Birders who observe rare birds should, whenever possible, verify their observations by video, photographs or tape-recorded vocalisations, to assist the Rare Bird Committee in subsequently assessing the report. However, in the absence of such documentation, a thorough written description – taken in the field, not from a field guide – will often suffice. As detailed a description as possible should be written while observing the bird, or as soon as possible afterward, and always *before* consulting a field guide. A good description should commence with the size, shape, colour and patterns of the bird, preferably proceeding from the bill and head to the upperparts, wings, underparts, legs and tail, respectively. A quick drawing with labels is recommended. Information regarding viewing conditions should also be noted, including the date, time, weather, light conditions, distance, optical aids used, other species present for comparison, and other observers present. Descriptions should be written as candidly and honestly as possible. Thereafter, discrepancies between what was observed and what appears in a field guide (birds often vary somewhat from illustrations) could be added. An observer's reputation is an important factor, requiring caution, honesty and humility.

Observations of unusual birds should be promptly reported to the editor of the South-eastern Caribbean Bird Alert (SCBA), a monthly e-mail and website service sponsored by the Trinidad & Tobago Field Naturalists' Club. Details for birds on the review list of the Trinidad & Tobago Rare Bird Committee (TTRBC) should be submitted to the committee's secretary for formal review. Pertinent contacts and further information concerning these organisations can be found on the internet, although their addresses may change and require a web search. The current SCBA website is www.wow.net/ttfnc/rarebird.html and the current TTRBC website www.geocities.com/ttrbc

# ETHICS

When observing or photographing birds, avoid approaching them too closely, especially rare species or those that are nesting or roosting. Also avoid overuse of playback or other methods used to attract birds. The risk of predation, abandonment of a nest or dispersal to a new territory elsewhere increases when birds are disturbed repeatedly. Encounters with rare birds should be reported to local authorities and shared with other birders, but be discrete in sharing information about the locations of sensitive species that are valuable to non-birders as food (e.g. piping-guans, boobies) or cagebirds (e.g. seedeaters, finches).

Also respect the law and rights of everyone. Do not enter private property without the owner's permission, or damage private property. Be courteous and respectful with those you meet, to protect the reputation of birders and safeguard access to prime birding areas.

# SECURITY

By the very nature of the activity, birding often takes place well off the beaten track and, like everywhere else in the world, it is necessary to be both streetwise and aware of personal safety. Criminal attacks away from traditional tourist areas are very rare and on birders are almost non-existent. However, we urge both discretion and vigilance at all times.

There are four species of venomous snakes on Trinidad, but none on Tobago. When walking along narrow trails through the forest, pay as much attention to where you are walking, as to where you are watching. Neither Fer de Lance nor Bushmaster is inherently aggressive, but neither reacts benevolently to being trodden on! If you are fortunate to see one of these snakes in the vicinity of your intended path, give it a wide berth. Extra vigilance is necessary in March–April, the traditional mating period for pit vipers.

Safety also embraces personal comfort. Insect bites can be irritating at best, disabling at worst, and Trinidad & Tobago has a wealth of possibilities. One of the most annoying, frequently encountered in patches of short grass, are chiggers, a tiny red mite. Whilst not dangerous, prevention is far preferable to cure – long sleeves and long trousers tucked into socks offer reasonable protection.

# WHERE TO WATCH BIRDS ON TRINIDAD & TOBAGO

## THE NORTHERN RANGE

The Northern Range is actually the north-easternmost extreme of the Andes chain and extends, as a band of forested hills, right across the northern Caribbean coastline of Trinidad. The two highest peaks, Cerro del Aripo and El Tucuche, reach nearly 950 m above sea level. The forest sports a bird species list of well over 200, and productive birding is to be found off the Blanchisseuse Road, the only vehicular highway that bisects the range at all elevations. A large number of hunting trails branch off from the road. Such trails can transect many miles and, whilst they may lead to secretive species, it is extremely easy to get lost without navigational aid. Birders must also be aware of the potential danger from the two highly venomous pit vipers present in this area.

### Asa Wright Nature Centre

Situated 17 km along the Blanchiusseuse Road from the main east–west highway at an altitude of c.360 m, Asa Wright comprises approximately 80 ha of coffee and citrus estate within seasonal rainforest. The estate boasts a bird list in excess of 160 species and is the easiest place on the island to find Channel-billed Toucan, Blue-crowned Motmot, Bearded Bellbird and Oilbird. From the balcony, it is usual to find six species of hummingbird, including both Tufted Coquette and White-necked Jacobin, whilst all three hermits can be seen on the entrance track. One of the main forest trails holds leks for both White-bearded and Golden-headed Manakins, and it is quite possible to find all three trogons, together with both Golden-olive and Chestnut Woodpeckers, Great and Barred Antshrikes, Plain-brown and Cocoa Woodcreepers, and Cocoa and White-necked Thrushes. Other regularly seen species include Grey-fronted Dove, Blue Dacnis, all three honeycreepers, at least seven species of tanagers and many Crested Oropendolas. The forest harbours secretive species including Little Tinamou, Grey-throated Leaftosser, White-bellied Antbird and Black-faced Antthrush – all of which, however, are more often heard than seen.

### Morne Bleu

The Radio and Tropospheric Scatter Station at c.670 m marks the highest point along the Blanchisseuse Road that a vehicle can access the Northern Range. Once inside the gate, the track to the station compound has specialities including Scaled Pigeon, Grey-rumped, Band-rumped and Chestnut-collared Swifts, Blue-headed Parrot, Collared Trogon, Channel-billed Toucan, Dusky-capped Flycatcher, Hepatic and Speckled Tanagers, and Sooty Grassquit.

### Las Lapas Ridge and Trace

This is a level portion of the Blanchisseuse Road, running north from the base of the Morne Bleu track for 1–2 km to a sharp right-angled bend with a lookout. Where the road begins to descend, a broad track branches off to the left through the forest, eventually reaching the villages of La Pastora and Lopinot. This is one of the most reliable places in the Northern Range for Ornate Hawk-Eagle, although even here luck is required. More regular fare includes: Squirrel Cuckoo, all three hermits, all three trogons, Red-rumped and Golden-olive Woodpeckers, Streaked Xenops, Stripe-breasted Spinetail, Grey-throated Leaftosser, White-bellied Antbird, Black-faced Antthrush, Plain Antvireo, White-flanked Antwren, Slaty-capped, Dusky-capped and Euler's Flycatchers, Turquoise, Bay-headed and, at the appropriate season, Swallow Tanagers.

### Paria junction

By continuing north on the Blanchisseuse Rd, and descending a further 1 km, you reach Paria junction. Whilst this is not a prime site for specialty species, exploring the roadside forest in either direction often produces birds missed elsewhere.

### Morne la Croix

Halfway between Paria junction and the small town of Blanchisseuse is the village of Morne la Croix. About 1 km beyond the village, the road veers sharply right and there is a wide track to the left. An hour's birding here could produce Short-tailed, Zone-tailed and White Hawks, Blue-headed Parrot, Rufous-tailed Jacamar, Pale-breasted Spinetail and Yellow-rumped Cacique, whilst Southern Rough-winged Swallows often perch on the overhead wires.

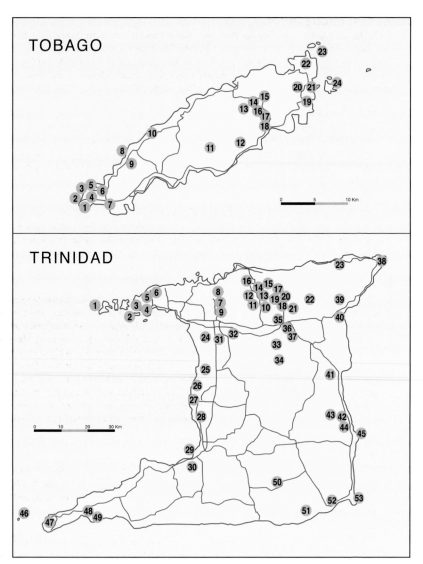

Major roads and recommended birding localities in Trinidad and Tobago. Bocas Islands: 1 = Chacachacare; 2 = Gaspar Grande. Northern Trinidad: 3 = Scotland Bay; 4 = Point Gourde; 5 = Morne Catherine; 6 = Tucker Valley; 7 = Maracas Waterfall; 8 = El Tucuche; 9 = Mt St Benedict; 10 = Simla; 11 = Asa Wright Nature Centre; 12 = La Laja South Trace; 13 = Morne Bleu; 14 = Las Lapas Road; 15 = Brasso Seco; 16 = Morne La Croix; 17 = Cerro del Aripo; 18 = heights of Aripo; 19 = Chaguaramal; 20 = Aripo Cave; 21 = Hollis Reservoir; 22 = Cumaca Cave; 23 = Grande Riviere. Western Trinidad: 24 = Caroni Swamp; 25 = Cacandee; 26 = Waterloo; 27 = Orange Valley; 28 = Pointe-a-Pierre Wildfowl Trust; 29 = San Fernando; 30 = South Oropouche Swamp. Central Trinidad: 31 = Caroni Rice Fields; 32 = Trincity Sewage Ponds; 33 = Arena Forest; 34 = Arena Reservoir; 35 = Aripo Livestock Station; 36 = Wallerfield; 37 = Aripo Savannas Scientific Reserve. Eastern Trinidad: 38 = Galera Point; 39 = Salybia Waterfall; 40 = Matura Beach; 41 = Brigand Hill; 42 = Nariva Swamp; 43 = Bush Bush; 44 = Kernahan; 45 = Point Radix. Southern Trinidad: 46 = Soldado Rock; 47 = Fullerton Swamp; 48 = Austin Road South; 49 = Los Blanquizales Lagoon; 50 = Cat's Hill Road; 51 = Trinity Hills; 52 = Guayaguayare; 53 = Galeota Point. Western Tobago: 1 = Crown Point; 2 = Store Bay; 3 = Pigeon Point; 4 = Bon Accord Sewage Ponds; 5 = Bon Accord Lagoon; 6 = Buccoo; 7 = Lowlands; 8 = Turtle Beach; 9 = Grafton Estate; 10 = Arnos Vale Bay. Central Tobago: 11 = Hillsborough Reservoir; 12 = Goldsborough Waterfall; 13 = Main Ridge Forest Reserve; 14 = Centre Hill Trail; 15 = Gilpin Trace; 16 = Spring Trail; 17 = Argyle River Trail; 18 = Argyle Waterfall. Eastern Tobago: 19 = Merchiston Road; 20 = Speyside Overlook; 21 = Speyside; 22 = Flagstaff Hill; 23 = St Giles Island; 24 = Little Tobago.

## Marianne River Estuary

From the Caribbean coastline at Blanchisseuse, the road continues east to the suspension bridge over the Marianne River. Whilst construction work traffic has diminished the value of this site, the adjacent riverine forest is still one of the easiest places on the island to find Silvered Antbird. Other species occasionally found here, especially in the early morning, include Grey-necked Wood Rail, Green Kingfisher, Crimson-crested Woodpecker and Black-tailed Tityra.

## NORTH-WEST TRINIDAD

The north-western peninsula of Trinidad incorporates a large protected area of mixed forest at Chaguaramas, whilst the coastal road from Port of Spain eventually accesses Chaguaramas National Park. The forest here harbours a good mix of species, ideal for birders based in the capital as an introduction to Trinidad forest birding. It is quite possible to see up to 70 species in a single early-morning visit

### Morne Catherine

This is roadside birding along a private road climbing steeply through seasonal forest. To reach the top requires a 2–3-hour hike. Birding is best at the bottom and the top; the middle section can be very quiet. Whilst, or perhaps because, there is very restricted vehicular traffic on this road, be cautious. There are many sharp bends; vehicles will not expect anyone to be on the road, and they are driven accordingly!

### Chacachacare

Off the north-west tip of Trinidad lie a series of islands, known locally as the Bocas; the largest of these, Chacachacare, is easily accessible by boat. From the main jetty, a single winding road climbs steeply through seasonal forest for c.2 hours to reach the lighthouse looking out over the Venezuelan Paria Peninsula. Resident specialties include Blue-tailed Emerald, White-fringed Antwren, Mouse-coloured Tyrannulet, Northern Scrub Flycatcher, Bran-coloured Flycatcher, Fuscous Flycatcher and Streaked Saltator; all species that are either absent from or difficult to find on mainland Trinidad. Both Tropical Parula and Brown-crested Flycatcher are particularly abundant here. Additionally, during August–October, it is possible to find Yellow-bellied Seedeater and, if you are extremely fortunate, Lesson's Seedeater in the grasses just below the lighthouse. Chacachacare's proximity to the Venezuelan mainland offers real incentive to rarity finders. A word of caution, there is no fresh water on the island and the walk will be very hot.

## WESTERN TRINIDAD

A significant proportion of the western coastline facing the Gulf of Paria comprises mangrove swamp and tidal mudflats. Some of the mangrove closest to Port of Spain is now, sadly, decidedly unsafe to visit due to high crime levels. However, several sites are still frequently visited by birders and have hosted a number of major rarities over the years.

### Caroni swamp

This area comprises c.65 km² of brackish water and mangrove swamp that is home to the country's major heronry. Accessed from the north–south highway, 2–3 km south of the Grand Bazaar intersection, several boat operators provide daily tours into the swamp, departing at 4:00pm to witness the Scarlet Ibis roost, which occasionally numbers several thousand birds. Additionally, hundreds of Snowy Egrets, Little Blue and Tricoloured Herons, together with a few Neotropic Cormorants and Yellow-crowned Night Herons, come into roost. Other species regularly encountered whilst navigating the maze of mangrove channels include: Blue-winged Teal, Great Egret, Cocoi Heron, Clapper Rail, Common Potoo, Green-throated Mango, Green, Ringed and Pygmy Kingfishers, Straight-billed Woodcreeper, Black-crested Antshrike, Bicoloured Conebill and Red-capped Cardinal. However, you will need extremely good fortune to chance upon Boat-billed Heron, Rufous-necked Wood Rail or Mangrove Cuckoo, all of which are present but shy and wary. Note that when the ibis are nesting (May–September) the breeding birds do not use the communal roosts, and most birds are just seen flying over in small groups towards the breeding colony. Along the south side of the main waterway (known aesthetically as the Blue River, but more commonly as No. 9 Drain), a tarmac road leads to a newly built interpretive centre. Whilst the main value of the centre is its clean washrooms, the adjacent mangrove is a good area for Clapper Rail.

## Brickfields, the Waterloo floating temple, Orange Valley and Carli Bay

South of the mangrove, the area generically called Waterloo has extensive tidal mudflats. During the appropriate seasons, on a rising tide, large roosts of Laughing Gull, Large-billed and Yellow-billed Terns, and Black Skimmer occur, together with smaller numbers of Royal and Common Terns. The exposed mud provides feeding for Semipalmated and Collared Plovers, Black-necked Stilt and Southern Lapwing, Short-billed Dowitcher, Greater and Lesser Yellowlegs, and Western and Semipalmated Sandpipers. A number of major rarities have occurred here including Maguari Stork, Black-tailed Godwit, Terek Sandpiper, and Kelp, Franklin's and Sabine's Gulls. Access is from the Freeport exit of the highway, some 16 km south of Caroni swamp.

At the top end of the mudflats lies Brickfields settlement, *c.*2 km north of the Waterloo floating temple. Here, you are normally closest to the high-tide roost. Alternatively, good birding can be had from the temple and cremation site car park. Another well worthwhile area is the concrete jetty at Orange Valley, several kilometres to the south. The final birding spot in this area is a further 8–9 km south, immediately before entering the industrial complex at Port Lisas. Known locally as Carli Bay, a small area of grassland just before the beachfront holds good numbers of Saffron Finch. Both Rufous Crab Hawk and, at low tide, Greater Flamingo sometimes occur but both are rare. A modicum of caution is required at this site; there have been isolated crime incidents – never stray too far from your vehicle.

## SOUTH-WEST TRINIDAD

Sadly, and inexplicably, relatively little birding is done south of San Fernando. The extensive mangrove swamp at Oropouche, freshwater marshes at the Pitch Lake and the entire Icacos peninsula are all undoubtedly seriously under-watched. Most resident and visiting birders live or stay in northern Trinidad. Birding the south is very time-consuming, as the main problem is one of access. The road infrastructure is rustic to say the least and padlocked gates, which necessitate a long hot walk to get in, control the access tracks into Oropouche. The Pitch Lake attracts a number of undesirables posing as guides and it takes a long time to reach Icacos. However, many lowland wetland specialties can be found at one site.

### Sudama steps

This is open-country birding, either side of a riverbank, looking out onto freshwater marsh on one side and a mangrove hedge on the other. Residents in the area include Pinnated Bittern, Long-winged Harrier, Greater Ani, Green Kingfisher, Spotted Tody-Flycatcher and Masked Yellowthroat. Additionally, wintering Prothonotary and Yellow Warblers, American Redstart and Northern Waterthrush can be found in the mangrove.

Sudama Steps is a Hindu open stage for religious worship, some 7 km south of San Fernando on the left-hand side of Pluck Road, immediately after a narrow bridge. There is a large car park. The best birding is to be had by crossing the road and walking west along the south bank of the river.

## NORTH-EAST TRINIDAD

The sparsely populated north-east of the island is far removed from the hustle and bustle of the east–west corridor. Extensive areas of relatively unspoilt lowland rainforest remain.

### Galera Point

Having driven the coastal road north, the north-easternmost point of Trinidad is marked by Toco lighthouse, where the Caribbean meets the Atlantic. In late April–late July, both Bridled and Sooty Terns occasionally feed just offshore. From August–October, this is the most accessible point to study seabird migration, with large numbers of Brown Noddy and Common Tern, together with smaller numbers of Roseate and Sandwich Terns, passing east. The stunted *Clusea* trees in the car park can be the first landfall for Nearctic passerine migrants, and both Blackpoll Warbler and Yellow-billed Cuckoo have been found on several occasions. It is also one of the most regular sites to find perched Lilac-tailed Parrotlets.

### Montevideo Trace

By continuing on the coast road west, you eventually reach the village of Grande Rivière, famous for its nesting Giant Leatherback Turtles in April–August. A hike up Montevideo Trace reaches a hilltop and large clearing on the left-hand side. Here, amongst the nutmeg trees, Trinidad Piping-Guan can often be found in the early morning. Other species frequently seen, both from the clearing and further west towards the river, include Grey-headed, Plumbeous and Swallow-tailed Kites, Lesser Swallow-tailed Swift, Channel-billed Toucan, Crimson-crested Woodpecker, Silvered and White-bellied Antbirds, and Black-tailed Tityra.

## EASTERN TRINIDAD
Birding the east coast of Trinidad primarily focuses on the perimeter of the Nariva swamp, an extensive area of freshwater marsh, part of which is a Ramsar site.

### Manzanilla beach
Having driven east along Eastern Main Road, the Atlantic coast is reached at Manzanilla. Whilst not a major seawatching venue, it is possible to see Leach's Petrels close inshore during late February to late April. More records of Cory's Shearwater have come from Manzanilla than anywhere else on the island.

### Cocos Bay Road
From Manzanilla, the road follows the coast south through well over one million coconut palm trees, interspersed with narrow fringes of mangrove. Here, during the heat of the day, raptors seek the shade and a slow drive should produce most, if not all, of Common Black, Savanna, Zone-tailed and Grey-lined Hawks, Pearl and Plumbeous Kites, Yellow-headed Caracara and Osprey. The small stands of roadside mangrove are good for Green-rumped Parrotlet, Pygmy and Green Kingfishers, Red-rumped Woodpecker, Black-crested Antshrike and Silvered Antbird.

### Kernaham
After c.17 km there is access to Kernaham settlement. A series of rough tracks beyond the village crisscross this part of the swamp. Common birds here include Black-bellied Whistling Duck, American Purple Gallinule, Wattled Jacana, Orange-winged Parrot, Red-bellied Macaw, Yellow-chinned Spinetail, Pied Water and White-headed Marsh Tyrants, and Red-breasted and Yellow-hooded Blackbirds. Specialties of the area include Pinnated Bittern, Long-winged Harrier and Azure Gallinule.

### Bush Bush
Situated in the heart of Nariva swamp, and best accessed by boat, is Bush Bush forest. A permit is required to visit this site and several eco-tour operators in Trinidad provide tailor-made tours. Bird specialties of the area include Bat Falcon, Red-bellied Macaw, Yellow-crowned Parrot, Crimson-crested Woodpecker and Plain Antvireo. It is also the site of a re-introduction programme for Blue-and-yellow Macaw.

## CENTRAL TRINIDAD
Birding in this area combines lowland rainforest, secondary scrub and, especially, large tracts of wet savanna that hold a number of specialty species.

### Mount St Benedict
For birders based at the Pax Guest House, at the southern base of the Northern Range above the town of Tunapuna, the forest on Mount St Benedict provides an easy and safe introduction to Trinidad birding. Within walking distance of Pax, the area has a bird list of more than 130 species including White Hawk, Grey-headed Kite, Ferruginous Pygmy and Tropical Screech Owls, Lesser Swallow-tailed Swift, all three trogons, Bright-rumped Attila, and White-bearded and Golden-headed Manakins. From the grounds at Pax it is regularly possible to see up to 11 species of hummingbirds.

### Aripo Agriculture Research Station
This site is open to birders on weekdays, though permission should be obtained upon entry. Birding is strictly from the roadside as the station works with Water Buffalo, which are renowned for their unpredictable temperament. The site is accessed from Eastern Main Road, some 6 km east of Arima. Regularly seen species include Cocoi and Striated Herons, Savanna Hawk, Southern

Lapwing, Least, Spotted and Solitary Sandpipers, Green-rumped Parrotlet, Striped Cuckoo, Yellow-chinned Spinetail, Pied Water and White-headed Marsh Tyrants, White-winged Swallow and Red-breasted Blackbird. Specialties include Ruddy-breasted Seedeater and Grassland Yellow Finch, both of which are extremely difficult to find away from this site.

## Cumuto railway line

Aripo Savanna is a large protected area of vegetated wetland holding large stands of Moriche Palm. Several birding sites exist here; currently the easiest of access and the safest is to walk along a former railway line, immediately north of Cumuto village. Birds regularly recorded in the area include White Hawk, Bat Falcon, Red-bellied Macaw, Fork-tailed Palm Swift, Ruby Topaz, White-tailed Goldenthroat, Bran-coloured and Sulphury Flycatchers, Black-tailed Tityra and Yellow-rumped Cacique. There are two specialties here, both are rare residents and will require good fortune to find: Rufescent Tiger Heron and Moriche Oriole.

## JOURNEYING BETWEEN TRINIDAD AND TOBAGO

An inter-island ferry operates daily between the islands. The operation varies, depending on schedule, between a normal ferry, where it is possible to get out on deck and the journey takes up to five hours, and a high-speed catamaran, which is much faster with no outside access. Birding is sparse from the ferry and non-existent from the catamaran. The only stretch of water that occasionally holds birds of interest, and usually only in May–August, from the ferry is the final two hours. If you consider using the ferry, ensure that the relevant crossing is during daylight. All return crossings to Trinidad are currently made at night.

## SOUTH-WEST TOBAGO

Areas of open fresh water are few and far between in Tobago. All of the worthwhile areas of freshwater marsh and mangrove, and the best lowland forest are in this portion of the island.

## Bon Accord sewage ponds

Situated within reasonable walking distance of most of the tourist accommodation at this end of the island, between the main Milford Road and the coastal lagoon, this is a series of four rectangular roadside ponds. There is no unauthorised access and it is necessary to look through the fence. Nevertheless, checking the site either early morning or late afternoon can be very productive. Species regularly found in the appropriate season include Least Grebe, Anhinga, Great and Snowy Egrets, Little Blue, Tricoloured and Green Herons, both Yellow-crowned and Black-crowned Night Herons, Black-bellied Whistling Duck, Blue-winged Teal, White-cheeked Pintail, Spotted and Solitary Sandpipers, Greater and Lesser Yellowlegs, Belted Kingfisher, Grey Kingbird and Caribbean Martin. It is probably the most reliable site in the islands for Little Egret. Rarities in recent years have included Western Reef Heron, American Wigeon, Ring-necked Duck, Buff-breasted Sandpiper, Wilson's Phalarope and Cliff Swallow. Walking down the side of the mangrove to the bay occasionally produces Mangrove Cuckoo.

## Hilton grounds

For birders staying at the Hilton hotel, the golf course and grounds have several freshwater ponds and a boardwalk through a stand of mangrove. This is a particularly good site for Anhinga, Black-bellied Whistling Duck, Belted Kingfisher and Mangrove Cuckoo, and normally holds at least one rare duck per year. Recent finds have included Ring-necked Duck, Lesser Scaup and Northern Shoveler. Tobago Plantations also have some sewage ponds, adjacent to which can be very productive, but permission is necessary for access. Least Grebe, Blue-winged Teal, Solitary Sandpiper, and Greater and Lesser Yellowlegs are regular and, of greater interest, both Pied-billed Grebe and Sora have been found on several occasions.

## Grafton sanctuary

Accessed from Shirvan Road, the lowland dry forest at Grafton Sanctuary is open from dawn to dusk. Several easy trails branch into the woodland, yet the main track up the hill is often the most productive. Species regularly found include Rufous-vented Chachalaca, Pale-vented Pigeon, Eared Dove, Green-rumped Parrotlet, White-tailed Nightjar, Ruby Topaz, Blue-crowned Motmot, Rufous-tailed Jacamar, Red-crowned Woodpecker, Olivaceous and Cocoa Woodcreepers, White-fringed Antwren, Yellow-breasted, Fuscous and Brown-crested Flycatchers, and Blue-backed Manakin.

## Adventure Farm

Close to Arnos Vale hotel, Adventure Farm is worth a brief visit, if only to watch the action at the hummingbird feeders. Rufous-breasted Hermit, White-necked Jacobin, Black-throated Mango, Ruby Topaz and Copper-rumped Hummingbird are all regular visitors. A small entrance fee is levied.

## Crown Point

The extreme south-west tip of the island can be good for seabirds feeding close inshore, including any or all of Roseate, Royal, Sandwich, Cayenne and Sooty Terns, and Brown Noddy.

## TOBAGO RAINFOREST

The Main Ridge forest reserve, a large tract of lower montane rainforest, forms the backbone of the island and reaches c.580 m. A number of trails crisscross the ridge. Throughout the area, Great Black Hawks occasionally soar overhead, whilst Collared Trogon, Rufous-tailed Jacamar and, rarely, Venezuelan Flycatcher perch in roadside trees.

## Gilpin Trace

By far the most accessible site in Tobago to find the globally restricted White-tailed Sabrewing. Accessed from the Roxborough–Bloody Bay road, this winding trail leads though pristine forest besides streams and waterfalls. Other species regularly found include Red-rumped and Golden-olive Woodpeckers, Stripe-breasted Spinetail, Fuscous Flycatcher, Yellow-legged Thrush and Blue-backed Manakin. Both Olivaceous Woodcreeper and White-throated Spadebill occur but are much harder to find. Note, the trail can be exceedingly muddy. A local entrepreneur hires out rubber boots at the trailhead which are a worthwhile investment.

## NORTH-EAST TOBAGO

Though noted primarily for its seabird colonies, birders staying in one of the hotels at Speyside village can find many of the commoner Tobago species in the local seasonal and scrub forest.

## Merchison Road

Situated opposite the Speyside lookout on the Windward Road, this track winds through open second growth and small stands of seasonal forest. Species include both Great Black and Broad-winged Hawks, Blue-crowned Motmot, Rufous-tailed Jacamar, White-fringed Antwren, Ochre-bellied, Yellow-breasted, Fuscous and Brown-crested Flycatchers, White-winged Becard, Scrub Greenlet, Black-faced Grassquit and Giant Cowbird.

## Little Tobago

This heavily forested island is best accessed by boat, from one of the several tour operators based in the car park of Blue Waters Inn, Speyside village. Breeding species, which can easily be seen from the two lookouts, include Red-billed Tropicbird, Red-footed and Brown Boobies, Sooty Tern, Brown Noddy and Laughing Gull. With good fortune, a few Bridled Terns can occasionally be found. Whilst Audubon's Shearwater nests on the island, their strictly nocturnal habits mean that they are rarely seen from land.

## St Giles Rocks

These offshore islets lie at the convergence of the Caribbean and the Atlantic and hold very large numbers of Magnificent Frigatebird, and Brown and Red-footed Boobies, together with a small but stable population of Masked Booby. Access is by boat, from either Speyside or Charlotteville. There is no landing, so caution is paramount here; the waters surrounding the rocks can be exceedingly rough. Even if boatmen are prepared to take you out, pay serious attention to the swell. There is little value in being in the proximity of hundreds of seabirds if you are using both hands to hold on, and cannot raise your binoculars to your eyes!

## TINAMOUS – TINAMIDAE

Forty-seven species, all in the Neotropics, but just one occurs on Trinidad. They are small-headed, short-necked, plump-bodied, almost tail-less and short-legged, cryptic in coloration and wary in behaviour.

### Little Tinamou *Crypturellus soui andrei* 23cm

Adult has greyish-brown crown and face, pale grey throat and rest of plumage reddish-brown. Bill grey, legs grey-green. Female often larger and more richly coloured than male. Immature is duller and smaller. **Voice** A mournful tremulous series of whistles, rising in pitch and volume; a melancholy drawn-out *oooeeeuuu* with second syllable higher pitched. **SS** Unlikely to be confused. **Status** Common and widespread but shy and retiring resident of Trinidad's forests (absent Tobago). Walks quietly on forest floor. Stands upright when alarmed or alert.

## SCREAMERS – ANHIMIDAE

Three species, restricted to lowland freshwater marshes of South America, characterised by their large turkey-like shape with small head and bill, short thick neck, bulky body, a sharp spur at the bend of the short broad wings, short tail and long stout legs. One species formerly occurred on Trinidad.

### Horned Screamer *Anhima cornuta* 81cm

Adult has black-and-white scaled crown and neck, and long forehead quill. Face and throat black with golden-yellow eyes. Upper breast scaled black and white; lower breast brownish-black; belly and undertail-coverts white. Upperparts oily greenish-black with broad silver-white fringes to wing-coverts. Underwing-coverts white; flight feathers greenish-black. Bill and legs grey. Immature recalls adult. **Voice** A powerful, guttural *kuk kwoo kok kwow* **SS** Unmistakable. **Status** Former resident of E Trinidad freshwater marshes, now locally extirpated; no sightings since 1964 and no records for Tobago.

## CHACHALACAS AND GUANS – CRACIDAE

Fifty members of this arboreal, turkey-like family occur in the New World, from the SW USA to southern South America. Characterised by their small heads, long necks, sturdy legs, bare throat skin, short wings and long broad tails, chachalacas are comparatively small, slender and dull-coloured; guans are larger and darker.

### Rufous-vented Chachalaca *Ortalis ruficauda ruficauda* 55cm

Adult has grey crown and face with dark blue bare loral skin, reddish-brown eyes and bare red throat. Neck and upper breast grey-brown, becoming buff on lower breast and belly. Undertail-coverts brick-red. Mantle, wings and rump olive-brown. Tail dark iridescent grey-green, outer feathers broadly tipped rufous. Bill and legs grey. Immature recalls adult. **Voice** Far-carrying four-syllable *kaa kaa ri ko* or *kaow kaa ri ka*. **SS** Unmistakable. **Status** Abundant and widespread throughout Tobago, in dense to open forest, brush to gardens. Absent from Trinidad.

### Trinidad Piping Guan *Pipile pipile* 65cm

Adult has shaggy crown, broadly fringed and tipped white. Face, from behind eyes to basal half of bill, covered by bare pale blue skin; lower face-sides to throat and wattle bright dark blue. Rest of body, wings and tail sooty-black with metallic purple sheen and extensive white fringes to wing-coverts. Bill pale blue tipped black; legs pinkish-red. Immature recalls adult. **Voice** Very thin series of whistles, each rising in pitch. Whirrs wings when gliding. **SS** Unmistakable. **Status** Very localised and rare resident of Northern Range on Trinidad. Most easily seen in Grande Rivière area.

Little Tinamou

♀

♀

immature

♂

♂

Horned Screamer

Rufous-vented Chachalaca

Trinidad Piping Guan

# DUCKS AND GEESE – ANATIDAE

Some 159 species occur worldwide. Of the 17 species recorded in T&T, only two are resident on Trinidad, with a third on Tobago. Of the rest, five are non-breeding visitors from South America, the remainder visitors from the north. On Trinidad all ducks are very susceptible to the Oct–Feb hunting season. In flight, presence or absence of wing-stripes, pale inner forewings and rump patches are useful identification features.

### Fulvous Whistling Duck *Dendrocygna bicolor*                                   48cm

Large with obvious rump patch. Adult has dark brown crown. Face and underparts bright tawny-brown with narrow black eyeline, white neck-sides and flank barring, and white undertail-coverts. Neck brown with narrow central black stripe; mantle black, barred buff. Upperwing-coverts reddish-brown; flight feathers and entire underwing black. Rump and uppertail-coverts white; tail black. Bill and legs dark grey. Immature similar but has paler and duller upperparts and grey-brown uppertail-coverts. **Voice** High-pitched two-note whistle *kwee-ooo*. **SS** Unlikely to be confused. **Status** Uncommon visitor to inland wetlands on Trinidad; has bred. Most records Apr–Oct. Not seen on Tobago for over 30 years

### White-faced Whistling Duck *Dendrocygna viduata*                              43cm

Medium-sized with all-dark wings. Adult has white forehead and foreface; rear crown and face black. Throat white bordered by broad black band. Neck and breast chestnut. Central belly black, with black-and-white flanks barring. Upperparts black, broadly fringed brown; tail blackish. Upperwing has dark red inner wing-coverts, rest black. Underwing all black. Bill dark grey; legs pinkish. Immature duller with grey forecrown, face and throat. **Voice** A three-note whistle. **SS** Adult unmistakable. Immature from similar Black-bellied Whistling Duck by smaller size and absence of white upperwing-stripe. **Status** Scarce visitor to freshwater marshes and rice fields on Trinidad; most records Jun–Sep. Absent from Tobago.

### Black-bellied Whistling Duck *Dendrocygna autumnalis discolor*                49cm

Large and bulky with an obvious wing-stripe. Adult has brown crown. Face and throat pale grey with prominent white orbital and narrow black border. Neck reddish-brown; lower breast and breast-sides pale grey. Belly black, undertail-coverts patchily black and white. Mantle reddish-brown; rump and tail black. Upperwing pattern distinctive, with buff lesser coverts, pale grey median coverts and broad white wing-stripe on greater and primary coverts. Flight feathers and entire underwing black. Bill red; legs pink. Immature similar but lacks throat border, has buff-brown rather than reddish-brown breast and neck-sides, with rest of underparts patchily grey and black. Mantle and rump dull brown. Bill and legs grey. **Voice** A musical five-note whistled *pip pip wee wee do*. **SS** Adult unmistakable. Immature from White-faced Whistling Duck by much larger size and wing-stripe. **Status** Uncommon resident of inland wetlands on Trinidad; more common on SW Tobago.

### Comb Duck *Sarkidiornis melanotos sylvicola*                                  76cm

Very large and bulky with broad wings. Adult male has black crown. Face and neck white, heavily spotted black. Underparts whitish, with contrasting black flanks. Entire upperparts oily greenish-black; rump paler in some lights. Underwings black. Bill and legs black with large protruding knob of dark grey skin at base of bill. Adult female smaller and lacks distinctive 'knob'. Crown paler; face and neck have less pronounced speckling; flanks greyer. Immature has dark brown crown; face and all underparts buff with dark eyeline and dark flanks mottling. Entire upperparts dark brown. **Voice** Usually silent. **SS** Unlikely to be confused. **Status** Rare visitor to inland wetlands of W Trinidad, with just four records in last 12 years; noted Jun–Oct including party of 17, Aug 2000.

### Snow Goose *Chen caerulescens caerulescens/atlanticus*                        71cm

Plumage polymorphic. White morph adult entirely white with black outer flight feathers. Occasionally head is stained buff. Bill and legs pink. Blue morph adult has white head and upper neck, rest of underparts dark blue-grey and white undertail-coverts. Lower neck, mantle and rump dark blue-grey; tail paler. Wing-coverts grey, broadly fringed white; flight feathers much darker. Underwing-coverts white; flight feathers black. Immature white morph has grey-white forecrown, face and underparts. Rear crown and upperparts pale brownish-grey with white rump. Wing-coverts and inner flight feathers grey fringed white; outer pairs black. Bill and legs dark grey. Immature blue morph has dark grey-brown body plumage with paler undertail-coverts. **Voice** A nasal high-pitched bark. **SS** Unlikely to be confused. **Status** Accidental on Trinidad with no records for at least 20 years.

Fulvous Whistling Duck

White-faced Whistling Duck

adult

juvenile

Black-bellied Whistling Duck

adult

immature

Comb Duck

♂

♀

juvenile

Snow Goose

blue morph immature

white morph immature

white morph adult

blue morph adult

intermediate white/blue
morph adult

# PLATE 3: DABBLING DUCKS I

## American Wigeon *Anas americana* 48cm

Breeding male has creamy white forecrown; rest of crown, face and neck pale grey speckled black, with a broad, bottle green band over eye and ear-coverts. Breast and fore-flanks vinous pink; central belly and rear flanks white. Undertail-coverts black. Mantle brownish-pink, finely barred black. Upperwing sharply contrasting, with grey lesser coverts, bold white band on median coverts, green speculum and black flight feathers. Underwing grey with obvious white axillaries. Rump and uppertail-coverts grey-white; tail black. Bill pale blue, tipped black. Eclipse male lacks creamy forehead and green facial band. Head grey with large blackish eye-patch, contrasting sharply rufous-mottled breast and fore-flanks. Undertail-coverts barred black and white. Mantle and inner wing-coverts black with broad buff fringes; tertials black with broad white fringes. All other feathers as breeders. Adult female similar to eclipse male but has more orange tones to breast and flanks, and narrower white upperwing-coverts bar. Immature also similar but lacks white wing-coverts bar. **Voice** A 2–3-syllable whistle *wii wee weeu*. **SS** Adult male unmistakable. **Status** Very rare visitor to wetlands of SW Tobago, with just three records in last 12 years, all Jan–Feb.

## Blue-winged Teal *Anas discors* 39cm

Breeding male has blue-grey head with obvious white foreface crescent. Neck, breast and belly reddish-brown, spotted and barred black; rear flanks white, undertail-coverts and undertail black. Upperparts dark grey-brown, fringed buff. Inner forewing pale blue. Green speculum bordered white on inner secondaries; outer secondaries and primaries dark brown. Underwing white with dark grey leading edge and flight feathers. Bill dark grey. Eclipse male has dark brown crown. Face grey with black eye-stripe, white orbital and vague pale foreface crescent. Underparts mainly grey-brown with darker mottling but central belly paler; upperparts dark brown with buff fringes. Immature and adult female recall eclipse male but white loral spot replaces crescent and they lack white speculum border. **Voice** A thin coarse *quack*. **SS** Breeding male unmistakable. Other plumages from Green-winged Teal by overall grey-brown tone to body plumage, blue inner forewing, white loral area and darker bill. **Status** Common non-breeding visitor to coastal and inland wetlands on both islands. Most records Sep–May.

## Northern Shoveler *Anas clypeata* 48cm

Breeding male has bottle green head, becoming distinctly blackish on foreface and throat. Breast and upper flanks white; belly and lower flanks rufous. Undertail-coverts black. Mantle and rump black; scapulars have white fringes. Inner forewing pale blue bordered white, green speculum and primaries dark grey. Underwing-coverts white; flight feathers dark grey. Tail black with white sides. Bill dark grey. Eclipse male similar but duller with grey-green head and faint white loral crescent. White breast, rufous belly and dirty white undertail-coverts all heavily scalloped black. Mantle and scapulars fringed buff. Bill grey with orange sides and base; usually held downwards. Adult female and immature have buff crown and face streaked black and indistinct black eyeline. Underparts dark brown, broadly fringed buff. Mantle, rump and central rectrices black with brown subterminal bars and buff fringes; outer feathers pale grey. Inner wing-coverts blue-grey; speculum green bordered white; primaries dark grey. **Voice** Usually silent. **SS** Male unmistakable. Immature and female from similar Mallard by bill shape, speculum colour and greater underwing contrast. **Status** Accidental to inland wetlands on both islands, but in last 12 years only one record, of two birds, at Lowlands, Tobago, Feb 2003.

## Mallard *Anas platyrhynchos* 58cm

Breeding male has dark bottle green head and neck, bordered by narrow white throat ring. Breast reddish-brown; flanks and belly pale grey-buff; undertail black. Upperparts and wing-coverts grey; speculum blue fringed black and bordered by white stripe. Primaries darker grey. Underwing-coverts white; flight feathers grey. Central rectrices black; outer feathers white. Bill bright yellow. Eclipse male resembles female, but retains yellow bill and is rustier on breast. Adult female and immature have dark brown crown and grey-brown face with long narrow black eyeline. Throat whitish; rest of underparts warm brown with darker feather centres, paler on belly and buff undertail-coverts is spotted brown. Upperparts and wings as eclipse male. Bill orange with blackish markings. **Voice** A hoarse *quack* often in slightly descending series and a nasal *rhaeb*. **SS** Breeding male unmistakable. Other plumages from similar dabbling ducks by size, bill and speculum colour. **Status** Uncertain. Occasional sightings suggest it is perhaps a very rare non-breeding visitor to both islands, but some farmers keep feral birds.

American Wigeon

♀

♂

♀

♂

Blue-winged Teal

♀

♂
eclipse

♂
breeding

♀

♂

Northern Shoveler

♀

♂

♀

♂

♀

♂

Mallard

♂

## Muscovy Duck *Cairina moschata* 84cm

Very large and goose-like. Adult male has bushy-crested crown and red facial skin wattles. Entire plumage glossy greenish-black with conspicuous white inner forewing and underwing-coverts. Bill mainly black with flesh-pink patches; legs black. Adult female much smaller and duller black, and lacks wattles. Immature recalls adult female but lacks or has very restricted white in upper- and underwings. **Voice** Usually silent. **SS** Wild birds unmistakable, but many farmers have domesticated birds which vary in plumage though mainly piebald with warty faces, reddish bill and legs. **Status** Accidental at Trinidad wetlands: whilst at least one sighting has been reported, no documented records for many years.

## Northern Pintail *Anas acuta* 53–66cm

Breeding male has dark chocolate-brown head, neck and throat, white neck-sides stripe, breast and belly, and grey flanks. Undertail-coverts cream; undertail black. Mantle and rump grey with black-and-white elongated tertials. Upperwing grey with bronze-green speculum edged black with a buff inner border, and white trailing edge to secondaries. Elongated central rectrices black, rest grey. Bill grey. Eclipse male has buff-brown head, neck paler buff and rest of underparts brownish-grey with white or buff fringes. Mantle and rump grey; tail blackish but lacks elongated rectrices. Adult female and immature have plain brownish-fawn head. Underparts grey-buff with black chevrons and paler belly. Upperparts dark brown broadly fringed buff. Upperwing grey-brown with subdued brown speculum and white trailing edge to secondaries. Underwing plain grey. **Voice** A whistled *kwee*. **SS** Adult male unmistakable. Other plumages from similar dabbling ducks by shape and size, upper- and underwing colour. **Status** Very rare visitor to inland wetlands of both islands, with just five birds in last 12 years; Nov–Dec and Jun.

## White-cheeked Pintail *Anas bahamensis bahamensis* 43cm

Adult has brown crown and hindneck to eye level; lower face and throat white. Underparts buff-brown heavily spotted darker, with paler plain undertail-coverts. Upperparts dark brown broadly fringed buff. Uppertail pointed and buff-white. Inner flight feathers buff with broad green speculum; outer feathers dusky. Underwing dark brown with white axillaries and broad central stripe. Bill basally red, distally blue-grey. Legs dark grey. Immature similar to adult but duller. **Voice** Usually silent. **SS** Unmistakable. **Status** Uncommon resident at inland wetlands on Tobago and a breeding visitor to Trinidad with most records early May–late Oct.

## Masked Duck *Nomonyx dominicus* 34cm

Small, compact diving duck with flat crown, stiff erectile tail and obvious wing-patch. Sits very low in water, often just head and tail are visible. Breeding male has black forecrown and face; rear crown, neck, throat and breast brick-red. Flanks reddish-buff densely spotted black; belly to undertail buff. Flanks and upperparts brick-red scalloped black. Tail black. Upperwings black with prominent square white patch on inner flight feathers. Underwings dark grey with white axillaries. Bill bright blue with black tip. Non-breeding male has dark brown crown. Face creamy buff boldly marked by black eyeline and broader dark brown parallel line below eye. Adult female and immature similar to non-breeding male but have creamier faces with narrower dark brown facial lines, overall greyer body plumage and duller greyish bills. **Voice** Varied hissing and clucking notes. **SS** Unlikely to be confused. **Status** Rare and extremely shy resident of vegetation-covered freshwater wetlands on Trinidad, but only a rare visitor to Tobago.

## Green-winged Teal *Anas carolinensis* 37cm

Breeding male has chestnut crown and face with large tear-shaped bottle green band running back from eye, covering ear-coverts and lower face. Central breast yellowish-buff spotted black; central belly white. Breast-sides and flanks grey, faintly and finely barred black with an obvious vertical white breast-sides stripe. Undertail-coverts and undertail yellowish-buff, bordered black. Bill dark grey. Eclipse male has dark grey crown. Face grey-brown with indistinct black eye-stripe. Rest of body feathers dark brown broadly fringed buff-brown, with paler belly and undertail-coverts. Bill basally yellow, distally grey. Immature and adult female similar to eclipse male. **Voice** A clear ringing whistle. **SS** Adult male unmistakable. Other plumages from Blue-winged Teal by browner plumage, lack of pale loral area, bill size and colour, and dark grey inner forewing. **Status** Accidental to inland wetlands of SW Tobago, with just one documented record in last 12 years, in Nov 1998. A couple of undocumented reports from Trinidad.

Muscovy Duck

♀

♂

feral Muscovy x domesticated
white duck (variable)

Northern Pintail

♀

♂
breeding

Masked Duck

White-cheeked Pintail

♀

♂

♂
non-breeding

♀

♀

♀

♂

♂

Green-winged Teal ♂

### Ring-necked Duck *Aythya collaris* 43 cm

Medium-sized diving duck with steep crown peaking well behind eye and long bill. Breeding male has black head with golden-yellow eyes. Neck and breast black; belly grey-white with obvious white fore-flanks bar separating grey flanks; undertail-coverts black. Hindneck, mantle, rump and tail black. Upperwing-coverts black; flight feathers grey. Underwing-coverts white; flight feathers grey. Bill has black tip and white subterminal band; rest grey with thin white basal border. Eclipse male duller with brown tone to black, whitish loral patch and dusky-brown flanks. Adult female has blackish crown. Face grey with diffuse white loral crescent, white orbital ring and thin white postocular streak. Breast brown; belly grey-white. Flanks grey-brown. Upperparts brownish-black. Bill pattern duller than male. Immature recalls adult female. **Voice** Usually silent. **SS** From Lesser Scaup by crown shape and bill markings. Male additionally by flanks and mantle coloration; female by facial markings and absence of white at base of bill. **Status** Rare visitor to inland wetlands of Tobago with 12 birds in last 12 years, all Nov–Feb. No records from Trinidad since 1967.

### Lesser Scaup *Aythya affinis* 42 cm

Medium-sized diving duck with angular forehead, crown peaking behind eye and obvious wing-stripe. Breeding male has black crown and face, glossed purple in good light, with golden-yellow eyes. Neck and breast black; belly and flanks white; undertail-coverts black. Mantle grey-white finely vermiculated black; wing-coverts black. Flight feathers have black trailing edge and tips; secondaries white; inner primaries grey, outer feathers black. Underwing-coverts white; flight feathers grey. Rump and tail black. Bill pale blue with black nail. Eclipse male has brown head and breast; dirty brown-white belly and flanks. Mantle and wing-coverts darker and drabber. Adult female mainly brown, darker on upperparts with broad white ring surrounding base of bill, dark amber eyes and white belly. Upper- and underwings as male. Immature recalls adult female but lacks white band. **Voice** Usually silent. **SS** Unlikely to be confused. **Status** Rare visitor to inland wetlands on both islands, with just five records in last 12 years, all Dec–Feb.

### Southern Pochard *Netta erythrophthalma erythrophthalma* 49 cm

Large diving duck with peaked rear crown, rather long bill and obvious wing-stripe. Adult male has head, neck and breast black, with reddish eyes. Belly, flanks and undertail-coverts dark brown. Mantle, rump and tail black. Upperwing-coverts black; flight feathers white with a black trailing edge and wing-tips. Underwing-coverts black; flight feathers grey. Bill pale blue-grey. Adult female has brown head and creamy white band at base of bill, white throat and ear-coverts patch. Neck, breast and belly brown mottled darker; undertail-coverts creamy white. Upperparts dark sooty-brown. Upper- and underwings as adult male. Immature recalls adult female but has restricted white facial markings. **Voice** Usually silent. **SS** Adults unlikely to be confused. Immature from female Lesser Scaup by larger size, dark belly, white undertail-coverts and longer, broader upperwing-stripe. **Status** Historically, occasional visitor to inland wetlands on Trinidad, but no records for over 150 years.

## Ring-necked Duck

♀

♂

♂
eclipse

juvenile

## Lesser Scaup

♀
non-breeding

breeding

♂
breeding

♂
non-breeding

## Southern Pochard

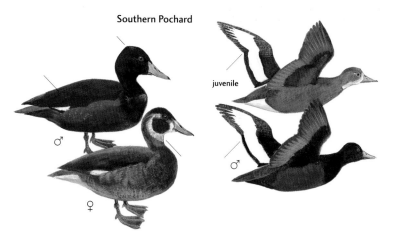

juvenile

♂

♂

♀

## GREBES – PODICIPEDIDAE

Nineteen species worldwide, but just two in T&T. They occur mostly on fresh water, frequently diving both for prey and in alarm. Flight rather weak.

### Least Grebe *Tachybaptus dominicus brachyrhynchus* 25cm

Tiny, with thin neck and thin, straight bill. Adult has black crown and face, with grey ear-coverts and yellow eyes visible at some distance. Throat grey-white; breast and belly brownish-grey with whiter flanks. Upperparts brownish-grey. In flight both upper- and underwing-coverts grey with white inner flight feathers. Bill black above, grey below. In breeding plumage, bill and throat black. Immature similar but paler, with a grey-and-white striped head. **Voice** A short, nasal *beep*. **SS** See Pied-billed Grebe. **Status** Uncommon resident that favours freshwater wetlands, especially sewage ponds.

### Pied-billed Grebe *Podilymbus podiceps antarcticus* 34cm

Rather large-headed, bulbous-billed and plump-bodied. Breeding adult has flat greyish-brown crown and face, and white eye-ring. Throat black, rest of underparts brownish-grey and white undertail-coverts. Upperparts greyish-brown; wings have indistinct white trailing edge. Bill ivory with thick black subterminal band. Non-breeding adult browner overall with unbanded yellowish-pink bill and grey-white throat. Immature has grey crown and grey stripes on reddish-brown face. **Voice** Male utters series of hoots, female low grunts. **SS** From Least Grebe by larger, plumper build, heavier bill, and plumage tones. **Status** Scarce resident of freshwater ponds and marshes on Trinidad, and a rare visitor to Tobago.

## PELICANS – PELECANIDAE

Frequently fly on slow deep wingbeats with long glides, and often in formation, with head drawn back. Just one species in T&T.

### Brown Pelican *Pelecanus occidentalis occidentalis* 122cm; WS 213cm

Non-breeding adult has creamy white crown and face. Neck white; rest of underparts brown-black. Upperparts silver-grey. In flight, upperwing contrast between silver-grey coverts and dark grey flight feathers. Bill brown with dark grey pouch and loral skin. Legs black, eyes orange. Breeder has creamy yellow crown, chocolate-brown neck with white stripe and bill and pouch tinged orange. Immature has grey-brown crown, face, neck and upper breast; below dirty brownish-white, above grey-brown with broad pale tips to coverts. Bill, pouch, loral skin and legs grey. **Voice** Usually silent. **SS** Unmistakable. **Status** Abundant on coasts of Trinidad, favouring sheltered waters, but less numerous on Tobago.

## ANHINGA – ANHINGIDAE

Two species worldwide, of which one occurs in T&T.

### Anhinga *Anhinga anhinga anhinga* 89cm; WS 114cm

Adult male has head, neck and underparts glossy-black. Upperparts black with buff terminal tail-band. Wing-coverts silver-grey, flight feathers black, underwing black. Bill pointed, orange-yellow; legs black. Adult female has buff-brown head, neck and upper breast. Immature has reduced silver-grey on wing-coverts. Only head and upper neck visible when swimming. **Voice** Clicks and croaks. **SS** Easily separated from Neotropic Cormorant by larger size; note flight jizz with long outstretched neck and bill, long fan-shaped tail and rapid wingbeats interspersed by long glides. **Status** Common year-round visitor to inland waters of both islands; has bred.

## CORMORANTS – PHALACROCORACIDAE

Thirty-nine species worldwide, only one in T&T. They are strong fliers and swimmers, but clumsy on land. When perched, often hangs wings out to dry. Swims low in water and surface-dives for prey.

### Neotropic Cormorant *Phalacrocorax brasilianus brasilianus* 66cm; WS 102cm

Adult has black head, neck and underparts with yellow gular patch and gape bordered by white V when breeding. Upperparts oily greenish-black with blacker flight feathers. Immature has dark brown head and underparts. Neck dark brown; rest of upperparts greenish-bronze. **Voice** Usually silent. **SS** Smaller than Anhinga with shorter neck and broader wings. **Status** Common year-round non-breeding visitor to inland and coastal waters of S and W Trinidad, with just one documented record for SW Tobago, in Nov 2004.

adult
non-breeding

adult
breeding

juvenile/
immature

Least Grebe

Pied-billed Grebe

adult breeding

juvenile/
immature

adult non-breeding

adult non-breeding

Brown Pelican

adult breeding

adult breeding
(when feeding young)

Anhinga

juvenile

♀

♂

adult
breeding

adult
non-breeding

immature

Neotropic Cormorant

juvenile

# PETRELS AND SHEARWATERS – PROCELLARIIDAE

Six species known from our coastal waters, all but Audubon's Shearwater extremely rare.

### Bulwer's Petrel *Bulweria bulwerii*      26cm; WS 70cm

Wings long and narrow, long tapered tail. Flight erratic, buoyant on bowed wings and deep wingbeats, gliding low over water. Sooty-brown with diffuse, pale buff carpal bar and uniform dark brown underwing. **Voice** Silent. **SS** See Sooty Shearwater. **Status** Accidental, Trinidad, Jan 1961.

### Cory's Shearwater *Calonectris diomedea borealis*      46cm; WS 117cm

Large, with long, slightly S-shaped wings. Soars and glides on bowed wings, arcing high above water. Sandy grey-brown crown and face, and white underparts. Uppertail-coverts have white tips; tail black. Underwing white with well-defined black leading and trailing edges. Bill yellow; legs pink. **Voice** Silent. **SS** From Great by lack of 'capped' effect, paler upperparts, less-obvious white uppertail-coverts, whiter underwing, all-white underparts and yellow bill colour. **Status** Rare passage migrant off E Trinidad and NE Tobago, Nov–May.

### Sooty Shearwater *Puffinus griseus*      48cm; WS 109cm

Long slender wings. Deep stiff wingbeats, alternating with long soaring arcs, low over water. Entirely sooty-brown plumage with contrasting silver underwing linings, often appearing as a white flash. Bill grey. **Voice** Silent. **SS** From Bulwer's Petrel by much larger size, longer wings, proportionately shorter tail, lack of upper-wing carpal bar, silver underwing-coverts, different flight action. **Status** Accidental, E Trinidad, Dec 1991.

### Manx Shearwater *Puffinus puffinus*      34cm; WS 84cm

Small to medium-sized; rapid, shallow wingbeats, shearing and turning low over water. Crown and upper face black; lower face slightly paler grey. Underparts white with dusky neck-side patches. Underwings white with ill-defined black trailing edge and wing-tip. **Voice** Silent. **SS** From most Audubon's with care, by all-white underparts including undertail-coverts. On water, longer wings project beyond tail. **Status** Very rare passage migrant off NE Trinidad; just two records, Feb 1997 and Oct 2002.

### Audubon's Shearwater *Puffinus lherminieri lherminieri*      31cm; WS 69cm

Small and short-winged. Quick fluttery wingbeats; less prone to gliding or soaring. Crown and upper face black; below white with dusky neck-side patches. Undertail-coverts usually dark grey. Note underwing pattern. **Voice** Nasal whines at burrows. **SS** From Manx by flight action and, usually, dusky undertail-coverts. **Status** Breeds off NE Tobago, including Little Tobago. Unless incubating, only approaches land after dark.

### Great Shearwater *Puffinus gravis*      48cm; WS 112cm

Large with straight, narrow wings. Stiff wingbeats; glides and soars in exaggerated arcs high over water. Black crown and face. Lower face to upper belly white with brown neck-sides. Underparts variably smudged dirty brown. Collar and uppertail-coverts white and tail black. Underwings white with ill-defined black leading and trailing edges, and black carpal bar. **Voice** Silent. **SS** See Cory's. **Status** Very rare passage migrant through T&T waters. Only four documented records in last 12 years, all sightings Jun–Jul.

# STORM-PETRELS – HYDROBATIDAE

Oceanic; identification based on wing shape, rump-patch shape and tail shape, and flight action.

### Wilson's Storm-Petrel *Oceanites oceanicus*      18cm; WS 40cm

Rather short, rounded wings with smoothly curved leading edge and straight trailing edge. Tail square-ended. Shallow, fluttering wingbeats. Sooty-black head and underparts with broad wrap-around white rump. Upperwing has greyish-buff carpal bar. Legs black, projecting slightly beyond tail; yellow webs between toes only visible very close. **Voice** Silent. **SS** From larger Leach's by shorter, straighter wings, tail shape, flight action and extent and shape of white on rump/undertail-coverts. **Status** Very rare passage migrant off E Trinidad and NE Tobago with no documented reports from land in last 20 years.

### Leach's Storm-Petrel *Oceanodroma leucorhoa*      20cm; WS 47cm

Long, angled and pointed wings, and rather long notched tail. Deep elastic wingbeats; frequently bounds, twists and turns, low over water. Sooty-brown head and underparts, with prominent buff carpal bar and narrow, V-shaped white rump. Underwing black; legs black. **Voice** Silent. **SS** See Wilson's. **Status** Scarce passage migrant off S and E Trinidad; rare off NE Tobago. Most records Jan–Apr.

**Bulwer's Petrel**

**Cory's Shearwater**

**Sooty Shearwater**

**Manx Shearwater**

**Audubon's Shearwater**

**Great Shearwater**

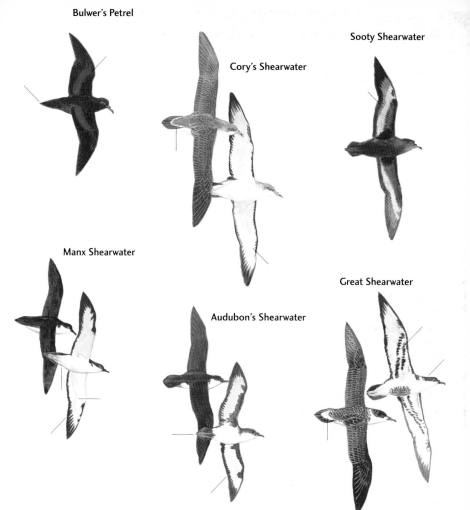

**Leach's Storm-Petrel**

**Wilson's Storm-Petrel**

## TROPICBIRDS – PHAETHONTIDAE

Three species occur throughout warmer oceans worldwide. Characterised by their graceful flight, elongated central tail feathers, principally white plumage with bold bill colours and upperwing markings. Two species occur in T&T; both on Tobago, one rarely to Trinidad coasts. Length measurements include central rectrices.

### White-tailed Tropicbird *Phaethon lepturus catesbyi*                    76cm; WS 94cm

Rather narrow wings and dainty jizz. Flies buoyantly with rapid wingbeats. Adult all white with short thin black eyeline, thick black upperwing carpal bar and black wedge on outer primaries. Bill usually yellow, sometimes orange. Immature lacks long tail feathers, and has thin black eyeline reaching ear-coverts and distinct black mantle, rump and wing-coverts barring. Bill yellowish with dark tip. **Voice** Silent at sea. **SS** Adult from Red-billed Tropicbird by smaller size, shorter wings, more buoyant flight, different upperwing and mantle markings, and bill colour. Immature with care by absence of black collar and on primary coverts, and more distinct mantle barring. **Status** At least one bird, possibly a pair, seen occasionally in recent years amongst Red-billed Tropicbirds on Little Tobago. Breeding possible.

### Red-billed Tropicbird *Phaethon aethereus mesonauta*                    102cm; WS 112cm

Long broad wings and extremely long central tail feathers. Flies gracefully, with deep purposeful wingbeats. Adult has white head with broad black line from bill to rear ear-coverts. Underparts white. Nape, mantle, rump and inner wing-coverts white with dense black barring and black wedge on outer primaries. Bill coral-red. Immature similar but lacks long rectrices; black eyelines meet on nape, forming a collar; primary coverts black and bill yellow. **Voice** Long shrill screaming *keeee keeee kee krrt krrt* around nest. Silent at sea. **SS** Adult from White-tailed Tropicbird by larger size, longer wings, bill colour, dense black barring on mantle and wing-coverts, and absence of black carpal bar. Immature by black collar, denser mantle barring and black primary coverts. **Status** Local but common resident on islands off E and NE Tobago, including Little Tobago. Rarer following post-breeding dispersal in Aug–Oct. Rare off Trinidad.

## FRIGATEBIRDS – FREGATIDAE

Five species occur worldwide, but only one occurs around T&T. In flight, appear very large with long, angular, pointed wings and long deeply forked tail. Wings arched and bowed. They soar effortlessly high over coasts and acquire food by piracy and with extremely agility, twisting and turning after prey. Never seen on water.

### Magnificent Frigatebird *Fregata magnificens*                    102cm; WS 229cm

Adult male has entire body plumage black with an inflatable red gular sac only visible briefly during breeding season. Bill dull pink; legs black. Adult female differs by contrasting white breast and lower neck-sides, buff upperwing carpal bar, and reddish legs. Immature has entire head and breast white, and similar carpal bar. Head colour varies, becoming progressively blacker with age. **Voice** Silent in flight and away from colonies. **SS** Unmistakable. **Status** Common resident in coastal waters of both islands, and abundant along windward coasts of Tobago. Occasionally seen high inland. Does not alight on water.

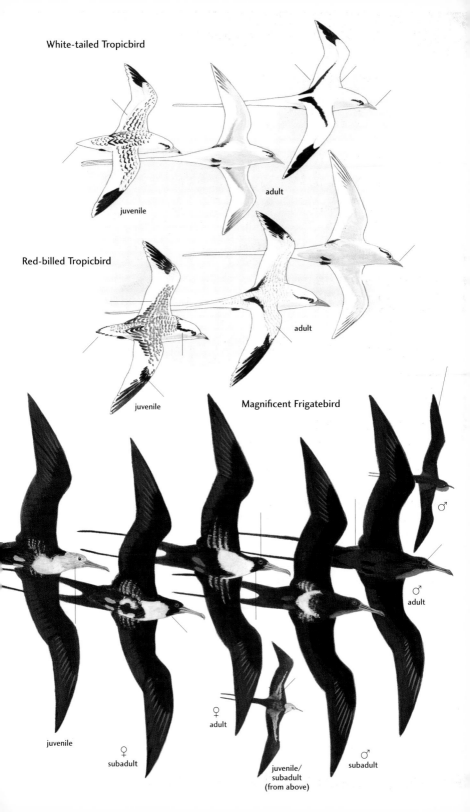

White-tailed Tropicbird

juvenile

adult

Red-billed Tropicbird

juvenile

adult

Magnificent Frigatebird

juvenile

♀
subadult

♀
adult

juvenile/
subadult
(from above)

♂
adult

♂
subadult

♂

# BOOBIES – SULIDAE

Ten species worldwide. Characterised by their conical bill, long neck, wings and tail, and spectacular plunge-diving feeding action. Fast, stiff, shallow wingbeats and short glides. Four species occur around T&T with the highest concentration of birds occurs off NE Tobago, whereas only two species are occasionally found off Trinidad coasts.

## Northern Gannet *Morus bassanus*                                          94cm; WS 183cm

Large and long-winged with long neck and pointed head. Adult has creamy white crown and face with pale eyes and long, thin black gape line. Underparts white. Hindneck creamy; mantle, rump and tail white. Wing-coverts and inner flight feathers white; outer feathers black. Bill pale greyish-yellow; legs black. Immature entirely dark grey-brown with faint but dense white spotting; white uppertail-coverts and pale underwing-coverts. Bill and legs dark grey. Feathers gradually whiten, attaining adult plumage in four years. **Voice** Usually silent. **SS** Combination of large size, creamy head and neck, black wing-tips, and white secondaries and tail separates adult from all boobies. Immature from similar Brown and Red-footed Boobies by larger size, longer wings, different bill colour, greyer, spotted plumage and white uppertail-coverts. **Status** Accidental off NE Tobago; adult Nov 1991.

## Masked Booby *Sula dactylatra dactylatra*                                 81cm; WS 158cm

Large and rather long-winged. Adult has white crown and face with black loral skin and lower cheeks. Rest of plumage white, except black flight feathers, trailing edge to underwing and tail. Bill yellow; legs pinkish-grey. Immature has dark grey-brown hood and white collar; breast, belly and undertail-coverts white; undertail-coverts dark grey. Mantle and upperwing-coverts brown, patchily spotted white. Rump white. Flight feathers and tail dark brown. Underwing white with dusky trailing edge, narrow central stripe and carpal patch. **Voice** Honks and whistles at colonies; otherwise silent. **SS** Adult from other boobies by white underwing; black upperwing flight feathers and tail. Immature by white underparts, neck collar and rump, yellow bill and orange feet. **Status** Local and uncommon resident on St Giles rocks off NE Tobago. Rare elsewhere.

## Brown Booby *Sula leucogaster leucogaster*                                76cm; WS 145cm

Bulky with rather short wings. Adult has dark brown head, neck and upper breast with yellow loral skin. Lower breast, belly and undertail-coverts white; undertail brown. Upperparts dark brown. Underwing-coverts white; flight feathers brown with broad white central stripe. Bill pinkish-yellow; legs yellow. Immature dull brown overall with slightly darker upperwing flight feathers and noticeably paler underwing-coverts. Bill grey; legs yellow. **Voice** Loud honking at colonies; otherwise silent. **SS** Adult unmistakable. Immature from Northern Gannet by lack of white uppertail-coverts and browner, unspotted plumage. Immature Red-footed Booby has more two-toned upperwing and darker underwing. **Status** Common resident in coastal waters of NE Tobago. Nests on rocks. Regularly seen along remaining coastline of Tobago and adjacent NE Trinidad. Decidedly uncommon elsewhere.

## Red-footed Booby *Sula sula sula*                                          71cm; WS 152cm

Comparatively small, slightly built and short-necked. Dimorphic. Brown morph adult mostly rich buff-brown with reddish loral skin. Rump and uppertail-coverts, and undertail-coverts creamy white. Wing-coverts brown; flight feathers black. Bill grey, legs red. White morph adult predominantly white, but secondaries have broad black trailing edge, primaries all black and underwing has black carpal patch. Immature drab grey-brown with contrasting brown upperwing-coverts and black flight feathers. Underwing all dark. Bill grey; legs dull orange-yellow. **Voice** Grunts and squawks at colonies; otherwise silent. **SS** White morph adult from larger Northern Gannet and Masked Booby by colour of tail and inner flight feathers. Brown morph adult from immature Brown Booby by strong upperwing contrast, paler underparts and creamy white tail. **Status** Locally common resident off NE Tobago. Nests in bushes and trees. Rare off of NE Trinidad.

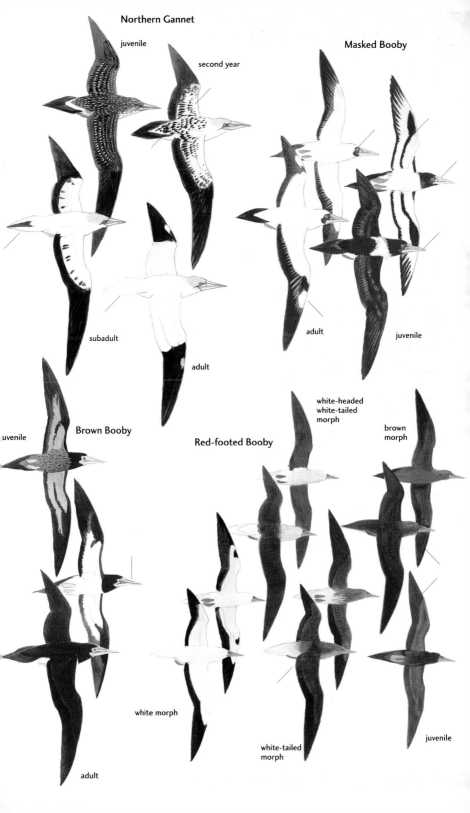

Northern Gannet

juvenile

second year

subadult

adult

Masked Booby

adult

juvenile

Brown Booby

juvenile

white morph

adult

Red-footed Booby

white-headed
white-tailed
morph

brown
morph

white-tailed
morph

juvenile

## HERONS – ARDEIDAE

A large and diverse family of 63 species worldwide. Vary in size and shape, but typically characterised by long neck and powerful pointed bill. A total of 22 species has been recorded in T&T, of which 15 have been found on both islands and seven only on Trinidad. Most are resident but some species' numbers are augmented by migrants from both North and South America.

### Rufescent Tiger Heron *Tigrisoma lineatum lineatum*                                                      71 cm

Appears large and 'hunched' at rest but long-necked in flight. Adult has vinous crown, face and neck with faint black crown barring, bright yellow orbital, eyes and loral skin. Throat and neck-sides white, breast striped rufous, buff and black. Belly and undertail-coverts buff with black-and-white flanks barring. Upperparts olive-green, faintly barred black. Bill dark grey with yellow base and underside of mandible; legs grey-green. Immature has bright orange-brown crown, face and neck-sides with black-barred crown and neck, and paler cream lores. Rest of underparts creamy white. Upperparts orange-brown, boldly barred black. Tail black with narrow white bars. **Voice** Quiet *urgh urgh urgh* and series of dull barks. **SS** Adult unmistakable. Immature from Pinnated Bittern by much brighter plumage with bolder black barring, orange-brown flight feathers and lack of darker crown and malar stripe. **Status** Rare resident in freshwater swamp-forest on Trinidad; breeding suspected. Most active dawn and dusk. Absent from Tobago.

### Agami Heron *Agamia agami*                                                      70cm

Large, long-necked and short-legged with exceptionally long thin bill. Adult has vinous crown and face, with pale blue nape plumes, green loral skin and reddish eyes. Rest of underparts vinous, with narrow white border to neck-sides and variable number of wispy silver-blue lower-neck plumes. Upperparts bottle green with vinous scapulars. Bill grey-green with yellow base; legs yellow. Immature has black crown; face and neck-sides brown. Rest of underparts buff-white, becoming white on lower belly and undertail-coverts, varyingly streaked brown, with narrow black border to neck-sides. Upperparts dark brown with dusky flight feathers. **Voice** Usually silent. **SS** Adult unmistakable. Immature from Tricoloured Heron by larger size, longer bill, dark underwing-coverts and habitat. **Status** Accidental to lowland forested streams of Trinidad.

### Pinnated Bittern *Botaurus pinnatus pinnatus*                                                      64cm

Usually adopts 'frozen' pose, with bill held nearly vertical. Adult has brown crown densely barred black. Face buff, finely barred brown and bordered by indistinct black malar stripe. Eyes and loral skin yellow. Throat white, rest of underparts whitish streaked buff. Nape buff densely barred brown. Upperparts buff streaked brown; flight feathers dusky. In flight, contrast between coverts and flight feathers pronounced. Bill and legs greenish-yellow. Immature paler with more subdued neck barring. **Voice** Occasional deep cough. **SS** From immature Rufescent Tiger Heron by paler buff plumage, streaked, not barred upperparts and contrast in upperwing. **Status** Uncommon and shy resident of freshwater marshes and rice fields on Trinidad. Usually solitary, keeping well inside dense cover. Absent from Tobago.

### Least Bittern *Ixobrychus exilis erythromelas*                                                      25cm

Tiny and richly coloured. Adult male has black crown, bright chestnut-buff face and neck-sides, and yellow loral skin. Rest of underparts white with buff breast streaking. Nape chestnut-buff, mantle, rump and tail black with thin buff-white scapular lines. Upperwing, shoulder and coverts bright buff, flight feathers dark rufous and underwing grey. Bill yellow with black culmen; legs yellow. Adult female similar but crown and mantle dark brown. Immature recalls adult female but has paler mantle and more extensive underparts streaking. **Voice** Loud *kick* and a guttural *churr*. **SS** From Stripe-backed Bittern by bright chestnut-buff neck-sides and face, solid dark upperparts, buff-white scapular lines and dark rufous flight feathers. **Status** Shy and uncommon resident of freshwater marshes on Trinidad. Absent from Tobago.

### Stripe-backed Bittern *Ixobrychus involucris*                                                      33cm

Small and rather pale. Adult has black crown, sandy-buff face and neck-sides, and yellow eyes and loral skin (latter red when breeding). Breast and belly paler buff, lightly streaked brown. Nape sandy-buff, mantle and rump boldly striped black, buff and white. Tail sandy-buff. Pale buff wing-coverts with rich rufous shoulder-patch and blackish flight feathers. Bill and legs yellow. Immature recalls adult. **Voice** An abrupt *hu*. **SS** From Least Bittern by paler sandy plumage, striped mantle and blackish flight feathers. **Status** Shy and uncommon resident of freshwater marshes and rice fields on Trinidad. Absent from Tobago.

Rufescent Tiger Heron

adult

immature

Agami Heron

juvenile

adult

Pinnated Bittern

Stripe-backed Bittern

Least Bittern

adult

juvenile

### Boat-billed Heron *Cochlearius cochlearius cochlearius*    58 cm

Short, squat and large-headed with unique bill shape. Adult has black crown, contrasting with white forehead and face, and large black eyes. Throat and breast white, belly and undertail-coverts chestnut-brown, with black rear flanks. Nape and foremantle black; rest of upperparts pale grey. In flight, prominent black underwing-coverts. Bill broad, black and often described as shovel-like. Legs greenish-yellow. Immature has dark brown crown, buff-brown face, throat and breast, becoming paler on belly and undertail-coverts; upperparts slightly darker brown. **Voice** Low-pitched guttural croaks. **SS** From Black-crowned Night Heron by smaller size, bill shape and dark underparts. **Status** Rare and rarely seen: a strictly nocturnal resident of dense mangrove swamps in W and SW Trinidad. Absent from Tobago.

### Black-crowned Night Heron *Nycticorax nycticorax hoactli*    64 cm

Stocky and large-headed with short neck, wings and tail. Adult has black crown, grey-white forehead and face, and red eyes. Underparts pale grey. Nape pale grey with several white plumes. Mantle and rump black; wings and tail pale grey. Bill black, legs greenish-yellow. Immature has grey-brown crown and face lightly streaked black. Throat, breast and belly brown with diffuse white streaking, undertail-coverts whitish. Nape, mantle and rump grey-brown. Wings and tail grey with bold white wing-coverts spotting. Upper mandible grey-green; lower dull yellow. Legs greenish-yellow. Feet project beyond tail in flight. **Voice** Monotonous loud *quaak*. **SS** Adult unmistakable. Immature from Yellow-crowned Night Heron by paler crown, longer bill with yellow lower mandible, blurred streaking on underparts and extensive bold white wing-coverts spotting. **Status** Common resident of mangrove swamps and adjacent freshwater marshes on Trinidad; both nocturnal and crepuscular. Less numerous on Tobago.

### Yellow-crowned Night Heron *Nyctanassa violacea cayennensis*    61 cm

Large-headed, thick-billed and slender. Adult has black crown and face with broad white cheek-patch, creamy central crown-stripe and several long white plumes on rear crown. Eyes red. Throat black, rest of under- and upperparts pale grey. Wing-coverts and tertials black fringed and tipped pale grey. Bill black, legs greenish-yellow. Immature has dark grey-brown crown and face, dull buff-brown underparts with narrow darker streaking on breast and belly, and grey-brown upperparts faintly spotted white; flight feathers darker grey. In flight feet and legs project beyond tail. **Voice** Hoarse *quark*, higher-pitched than Black-crowned. **SS** Adult unmistakable. Immature from Black-crowned Night Heron by more erect stance, darker crown, all-black stubby bill, longer neck, more distinct underparts streaking and less distinct wing-coverts spotting. **Status** Common resident on coasts of T&T, favouring mangroves and tidal mudflats; primarily nocturnal but often feeds during daytime on open mud.

### Striated Heron *Butorides striata striata*    46 cm

Adult has black crown and face to eye level. Lower face and neck-sides grey with yellow loral skin, yellow eyes and black gape. Throat, foreneck and breast white, with black breast streaking and rufous-buff breast-sides. Belly and undertail-coverts grey. Upperparts dark grey-green with pale fringes to wing-coverts. Upper mandible black, lower yellow. Legs orange-yellow. Immature has brown head and underparts with extensive buff and white streaks on neck and upper breast. Belly and undertail-coverts white. Upperparts grey-brown with extensive white spotting on wings. **Voice** Scolding *keeow*. **SS** Most adults from Green Heron by greyer face and neck-sides, but some are intermediate-looking (or hybrids) and may be inseparable in the field. **Status** Common resident of wetlands throughout Trinidad; rare on Tobago.

### Green Heron *Butorides virescens virescens*    46 cm

Adult has shaggy black crown and face to eye level. Lower face, throat and neck-sides deep vinous-chestnut with yellow eyes, greenish-yellow loral streak and white gape. Foreneck and breast creamy white, boldly striped chestnut; belly and undertail-coverts grey. Upperparts bottle green with buff fringes to wing-coverts and elongated blue-grey scapulars when breeding. Bill black above, mainly yellow below. Legs orange-yellow. Immature has dark brown crown. Face, neck and breast brown with extensive buff and white streaks. Belly and undertail-coverts white. Upperparts grey-brown with extensive white spotting on wings **Voice** Scolding *keeyow*. **SS** Most adults from Striated Heron by chestnut face and neck-sides. However, there is much variation and some appear intermediate, being possibly hybrids, and cannot be identified with certainty in the field. **Status** Common resident of inland and coastal wetlands on Tobago. Rare visitor to Trinidad.

**Boat-billed Heron**

juvenile

adult

**Yellow-crowned Night Heron**

adult

juvenile

**Black-crowned Night Heron**

juvenile

adult non-breeding

adult breeding

**Green Heron**

juvenile

adult

variant
(possible hybrid
Striated x Green)

**Striated Heron**

juvenile

adult

## Cattle Egret *Bubulcus ibis*                                                    51 cm

Rather small and rotund, being proportionately short-billed and thick-necked. Non-breeding adult all white. Bill and eyes yellow; legs (usually) greenish-yellow. When breeding, has an orange-buff stain to crown, mantle and breast. During courtship, bill and legs turn red. Immature initially has dark bill, lores and legs. Subsequently recalls non-breeding adult but has black legs. **Voice** Usually silent, very occasionally a hoarse croak. **SS** Unlikely to be confused. **Status** Abundant, widespread resident on both islands. The least aquatic egret, ubiquitous in habitat, often following grazing mammals.

## Little Egret *Egretta garzetta*                                                 63 cm

Adult all white usually with blue-grey lores and two head plumes, longer when breeding. Bill black with paler base to lower mandible, legs black with yellow feet which turn red during courtship. Immature recalls adult but has paler lower mandible. **Voice** Hoarse *arrh*. **SS** From Snowy Egret by sloping rather than rounded forehead, longer neck, loral skin colour and usually all-black legs. **Status** Scarce year-round resident in T&T; breeding unproven. Prefers wet rice fields and sewage ponds.

## Snowy Egret *Egretta thula*                                                     61 cm

Adult all white with yellow lores and eyes. When breeding, white frilly plumes adorn rear crown and tertials, and loral skin is briefly reddish. Bill black with greyish base to lower mandible. Legs black, often with greenish-yellow stain on back; feet golden-yellow. Immature recalls adult with even paler lower mandible. Very young birds can have greyish lores. **Voice** Hoarse *arrrh*. **SS** From Little Egret by slightly smaller size, shorter neck, slighter bill, frilly head plumes, more rounded crown, different loral skin and rear leg colour. From immature Little Blue Heron by bill shape, bill and leg colour and, except in youngest birds, by unmarked white body plumage. **Status** Abundant resident in wetlands across Trinidad; less numerous on Tobago. Numbers swelled by migrants from both North and South America.

## Little Blue Heron *Egretta caerulea*                                            61 cm

Adult has vinous crown, face and neck surrounding pale yellow eyes and grey loral skin. Remaining body feathers dark slate blue-grey. When breeding has extended scapular plumes; lores and legs darker. Bill basally pale grey, distally black; legs greyish-green. Immature initially white but increasingly is mottled slate blue-grey. Bill pale grey with black tip; legs dull pale green. **Voice** Croaks and squawks. **SS** Adult from Western Reef Heron by vinous head and neck, and lack of white throat. Immature from Snowy Egret by thicker, tapered bill, and bill and leg colour. From immature Reddish Egret by smaller size, bill shape and different bill and leg colour. **Status** Common resident in wetlands across T&T with numbers swelled by migrants from North and South America.

## Western Reef Heron *Egretta gularis*                                            61 cm

Adult entirely dark powdery grey, except white rectangle on chin, throat and lower cheeks, and pale grey undertail-coverts. Eyes bright yellow; bill dark grey with paler base to lower mandible and slightly drooping. Legs dark grey with greenish-yellow feet and white thighs. Immature has brown wash to body plumage with paler chin and throat. **Voice** Usually silent, but sometimes utters a guttural croak. **SS** From Little Blue Heron by conspicuous chin and throat patch, and different neck, bill and leg colours. White morph unrecorded in New World. **Status** Accidental: just two records, both in freshwater marshes: Trinidad, Jan 1986 and Tobago, Dec 2000–Jan 2002.

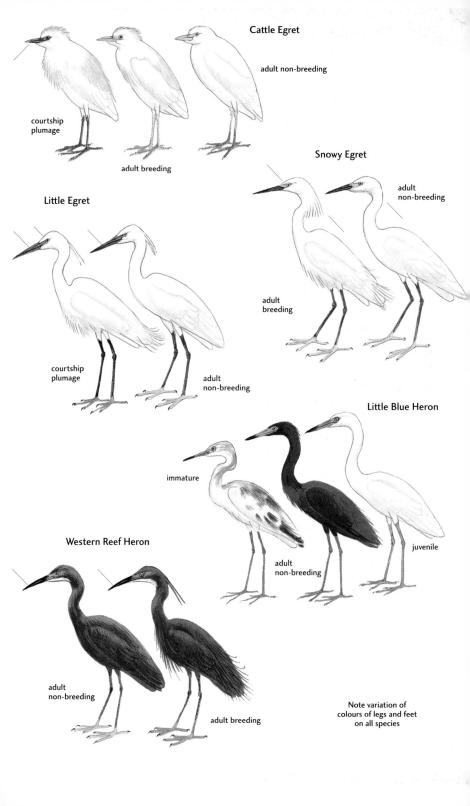

**Cattle Egret**

adult non-breeding

courtship
plumage

adult breeding

**Snowy Egret**

adult
non-breeding

**Little Egret**

adult
breeding

courtship
plumage

adult
non-breeding

**Little Blue Heron**

immature

juvenile

**Western Reef Heron**

adult
non-breeding

adult
non-breeding

adult breeding

Note variation of
colours of legs and feet
on all species

### Great Blue Heron *Ardea herodias* 117 cm

Adult has white crown bordered black. Face and throat white with yellow eyes and greenish-yellow loral skin. Neck vinous-pink to rufous with paler shaggy feathers at base. Foreneck and breast narrowly striped black and white. Belly dark grey, thighs rufous. Upperparts grey. Flight feathers darker with rufous inner leading edge and carpal 'headlights'. Bill yellowish, legs greenish-grey. Immature has sooty-grey crown and upper face, white throat, grey neck-sides mottled white, dirty white foreneck and rest of underparts with short black neck and breast streaking, and dull rufous thighs (white in some birds). Upperparts brownish-grey. Upperwing as adult. Upper mandible blackish, lower dirty yellow. **Voice** Deep *aark*. **SS** Adult unmistakable. Immature from Grey Heron by larger size, longer bill and neck, (usually) rufous thighs, inner leading edge and carpal, and darker grey upperparts. **Status** Uncommon non-breeding visitor to wetlands across T&T, though a few are present year-round.

### Great Egret *Ardea alba* 99 cm

Tall with sloping crown, substantial bill, snake-like neck and long legs. Deep heavy wingbeats. Adult has all-white body plumage, yellow-green loral skin, yellow eyes and thin black line from gape to behind eyes. When breeding, loral skin greener and has tertial plumes. Bill yellow, legs black. Immature recalls adult. **Voice** Deep *kraaak*. **SS** Unlikely to be confused. **Status** Common resident in wetlands across T&T with numbers swelled by migrants from North and South America.

### Cocoi Heron *Ardea cocoi* 121 cm

Adult has all-black crown and upper face, yellow eyes and blue-grey loral skin. Lower face, neck and breast white with short black streaks on foreneck and upper breast. Belly, undertail-coverts and thighs white with black flank patches. Upperparts pale grey. Wing-coverts even paler, contrasting with white carpal 'headlights' and black flight feathers. Bill yellowish-orange, with dark base to upper mandible; legs brownish-grey. Immature duller with pale grey-brown upperparts and greyish belly streaked buff. **Voice** Deep *curk*. **SS** Adult from Great Blue Heron by all-black crown, white underparts, especially thighs; black flank patch, paler wing-coverts, white carpal and darker flight feathers. From Grey Heron by larger size, longer bill, all-black crown, cleaner whiter underparts, black flank patch and paler wing-coverts. Immature from Great Blue by much paler appearance. Additionally from Grey Heron by larger size. **Status** Scarce non-breeding visitor to wetlands and pastures on Trinidad, with most records Dec–Jul. Rare on Tobago: just four records in last ten years.

### Grey Heron *Ardea cinerea* 94 cm

Adult has white crown, face and neck, with black crown-sides and short nape plumes. Underparts, including thighs, white with short black streaks on foreneck and breast. Upperparts grey, flight feathers black with white inner leading edge and carpal 'headlights'. Bill yellowish, legs grey. Immature duller with dark grey crown and much of face, paler forecrown, greyer underparts and greyish-yellow bill. **Voice** Harsh *kar-ark*. **SS** Adult from Cocoi Heron by smaller size, black restricted to head-sides, lack of black flanks and belly, and grey upperwing-coverts. From Great Blue Heron by smaller size, different neck, underparts and thigh colorations, paler grey upperparts and shorter bill. Immature from Great Blue by smaller size, shorter bill, different underparts and (usually) thigh colours. **Status** Very rare visitor to freshwater marshes and agricultural pastures in T&T, with five records in last eight years.

**Great Blue Heron**

white morph

adult breeding

courtship
plumage

adult
non-breeding

juvenile

adult

**Great Egret**

Cocoi Heron

courtship
plumage

adult breeding

adult
non-breeding

Grey Heron

### Reddish Egret *Egretta rufescens* 76cm

Dark morph adult has dull reddish crown, face and neck with shaggy rear crown and lower neck. Rest of plumage slate-grey. In flight has broad pale grey stripe across otherwise dark grey underwing. Bill distinctly two-toned; basally pink, distally black. Legs dark blue-grey. White morph adult is all white with bare parts as dark birds. Immature recalls adult, but dark morph duller and both morphs have blackish bill and legs. **Voice** Usually silent. **SS** Dark morph unmistakable. White morph from immature Little Blue Heron by bill and leg colours. **Status** Accidental in T&T with just one record in last 15 years; a dark morph on Tobago, Jul 2003. Forages alone in tenaciously held territory, constantly active, and often holds wings open when fishing.

### Purple Heron *Ardea purpurea* 90cm

Smaller and slighter than other large herons. Thin, snake-necked and thin-billed. Adult has black crown. Face and neck-sides ginger-brown with greenish-yellow loral skin, yellow eyes and black stripe across ear-coverts. Foreneck narrowly striped black and white, sides and lower breast brown with grey-and-white plumes. Belly and undertail-coverts streaked brown and black. Upperparts slate-blue. Lesser wing-coverts rich maroon-red, rest brownish-grey; flight feathers dark grey. Bill yellow with dark tip, legs basally yellowish, distally dark brown. Immature more subdued with dark grey crown; face and neck-sides bright buff with black ear-coverts stripe. Foreneck and breast-coverts streaked reddish-brown; rest of underparts buff-brown. Mantle and wing-coverts brownish-buff, fringed white. Flight feathers, rump and tail dark grey. **Voice** A gruff *krekk*. **SS** Unlikely to be confused. **Status** Accidental on Trinidad: juvenile in rice fields in centre of island, Sep 2002.

### Tricoloured Heron *Egretta tricolor tricolor* 66cm

Long thin neck and bill. Adult has dark blue-grey crown, face and neck-sides, with bright yellow loral skin. Foreneck and breast white streaked reddish-brown; rest of underparts white. Upperparts blue-grey with buff-brown tertial plumes and white rump. Bill orange with dark culmen; legs yellow-orange. Underwing-coverts white; flight feathers dark grey. When breeding has pair of long white plumes on nape and elongated reddish plumes on upper breast and mantle. Lores turn blue, bill basally blue and distally black, and legs red. Immature has reddish-brown crown, face and neck-sides, rest of underparts white. Fore mantle reddish, lower mantle, rump and tail blue-grey. Upperwing has reddish coverts and dark grey flight feathers; underwing as adult. **Voice** Usually silent. **SS** Adult unmistakable. Immature from Agami Heron by white belly and underwing-coverts; also habitat. **Status** Common resident, mainly in coastal wetlands, in T&T, with numbers swelled by migrants from North and South America.

Reddish Egret

juvenile

white morph

adult
non-breeding

adult
non-breeding

courtship
plumage

courtship
plumage

adult
breeding

Tricoloured Heron

Purple Heron

juvenile

adult breeding

adult
non-breeding

adult

juvenile

## IBISES and SPOONBILLS – THRESKIORNITHIDAE

Thirty-three species worldwide. Ibises have strongly decurved bills and rather short legs; they probe mud for crustaceans; Spoonbills have long, straight, spatula-shaped bills that they sweep from side to side in shallow water. Both fly with outstretched necks..

### Scarlet Ibis *Eudocimus ruber*     58cm

Adult entirely bright scarlet-red with black outer primary tips. Loral skin, strongly decurved bill and legs pinkish, though bill darker in breeding season. Immature has brown head, neck, upper breast and upperparts, and white lower breast, belly and rump. Subadult is grey rather than brown, and pink instead of white. Bill and legs pinkish-orange. **Voice** Usually silent away from the nest. **SS** Adult unmistakable. First-year inseparable in field from immature White Ibis. Thereafter, pink tones render it unmistakable. **Status** Locally abundant resident on mangrove-lined coasts of W Trinidad. Some movement to and from mainland South America. Very rare on Tobago with no records in last 15 years.

### White Ibis *Eudocimus albus*     58cm

Adult entirely white with black wing-tips. Facial skin, bill and legs bright red. First-year inseparable in field from immature Scarlet Ibis. Subadult has ever-increasing amounts of white in body plumage, rendering it unmistakable. **Voice** A nasal *urnk*. **SS** Adult unmistakable. **Status** Accidental to coastal mudflats and mangroves of Trinidad, with no documented records in last 20 years. No records on Tobago.

### Glossy Ibis *Plegadis falcinellus*     58cm

Breeding adult has dark reddish-brown head with metallic green gloss, dark eyes, greenish-grey lores edged with white. Body dark reddish-brown. Wings and tail metallic, appearing oily copper-bronze or green and black. Non-breeder has dark greyish-brown head and neck, densely flecked white. Rest of body dark brown with more subdued oily gloss to wings. Immature duller and darker still. **Voice** A nasal *urnn*. **SS** Adult from White-faced Ibis *P. chihi* (unrecorded in T&T) by lack of complete white border to bill base and red facial skin, dark not reddish eyes and brownish not bright red legs; immature by eye colour. **Status** Rare visitor to inland wetlands on Trinidad; only eight documented records in last 10 years. Accidental Tobago.

### Roseate Spoonbill *Platalea ajaja*     81cm

Adult has bare grey crown and facial skin. Neck white, rest of underparts soft pink. Mantle pale pink; wing-coverts brighter, with reddish shoulders. Flight feathers, rump and tail bright pink. Bill grey, legs reddish with grey 'knees' and feet. Immature similar but body mainly white, with soft pale pink wings. Bill yellowish-grey; legs grey. **Voice** A quiet grunt. **SS** Unmistakable. **Status** Very rare visitor to Trinidad with just two records from inland wetlands in last 10 years. Accidental on Tobago.

### Eurasian Spoonbill *Platalea leucorodia leucorodia*     82cm

Breeding adult has white crown and face with yellow orbital and loral skin. Chin yellowish; rest of underparts white with yellow breast stain. Upperparts white with yellowish bushy nape crest. Non-breeding adult all white. Bill black, tipped yellow; legs black. Immature all white with black wing-tips. Bill greyish-pink; legs dark grey. **Voice** Usually silent. **SS** At all ages, from Roseate Spoonbill by lack of pink tones, and adult by darker bill and immature by black wing-tips. **Status** Accidental on Tobago, an immature at Buccoo, Nov 1986.

## LIMPKIN – ARAMIDAE

Large marsh-dwelling bird, small-headed, long-necked and long-legged. Flies with neck outstretched, head and bill slightly drooped, on very broad wings with exaggerated deep wingbeats and well-splayed primaries.

### Limpkin *Aramus guarauna*     66cm

Chocolate-brown with greyer head and foreneck. Nape densely streaked white, with white scalloping on upper mantle and shoulders. Bill dull yellowish-grey with pale orange base. Immature recalls pale adult. **Voice** A loud drawn-out *kerauu* at dawn and dusk. **SS** From immature Scarlet and Glossy Ibises by bill shape, plumage colour, nape and mantle scalloping, and broader wings. **Status** Normally shy and secretive. Locally common resident in lowland wet savannas and marshes on Trinidad; absent Tobago.

Scarlet Ibis

juvenile

♀
adult

White Ibis

Glossy Ibis

juvenile

♂

adult
breeding

adult
non-breeding

Limpkin

Roseate Spoonbill

Eurasian Spoonbill

adult
non-breeding

adult

immature

juvenile

## STORKS – CICONIIDAE

Five species worldwide, one recorded on Trinidad. Characterised by large size, bulky bodies, heavy bills and long legs.

### Maguari Stork *Ciconia maguari*          100cm; WS 122cm

Large with long straight bill. Adult has white crown, face and underparts, occasionally stained yellow on throat; reddish loral skin and yellow eyes. Neck, mantle, inner wing-coverts and rump white; primary coverts, flight feathers and tail black. Bill dirty grey with reddish tip; legs grey-pink. Immature lacks red loral skin, with dark eyes and black wing-coverts spotting. **Voice** Usually silent. **SS** From all other storks by white head. **Status** Accidental to coast of W Trinidad; just one sighting, Sep 2001.

### Wood Stork *Mycteria americana*          102cm; WS 155cm

Large with long, thick slightly decurved bill. Adult all white with bare blackish skin covering head and neck. Flight feathers black. Bill and legs dark grey with pale feet. Immature similar but has yellow-brown stain to neck and yellowish-grey bill. **Voice** Usually silent. **SS** From much larger Jabiru by decurved bill, absence of red collar, black tail and black flight feathers on both wing surfaces. **Status** Accidental visitor to coastal wetlands of Trinidad with just one documented record in last 65 years.

### Jabiru *Jabiru mycteria*          132cm; WS 240cm

Huge, with very thick, slightly upturned bill. Adult all white, with bare black skin covering head and neck, occasionally a patch of white feathers on rear crown. Base of neck forms broad red collar. Bill and legs black. Immature pale grey with random white patches. **Voice** Usually silent. **SS** From much smaller Wood Stork by bill shape, red collar and all-white upper- and underwings. **Status** Very rare visitor to inland wetlands in T&T with just four records in last 20 years.

## FLAMINGOS – PHOENICOPTERIDAE

Five species worldwide. Characterised by extremely long neck and legs, and bill angled sharply downwards midway; one species recorded on Trinidad.

### Greater Flamingo *Phoenicopterus ruber ruber*          117cm; WS 152cm

Adult varies from bright salmon-pink to soft rosy-pink, with white loral skin and black flight feathers. Bill basally pink, distally black. Legs dull pink. Immature has pale brownish-grey crown and face. Neck greyish-white, rest of underparts paler, dirty white. Mantle, rump and wing-coverts grey with extensive white fringes. Flight feathers sooty-black. Bill grey with black tip; legs grey. Subadult has white body plumage with pink tinge. Flight feathers, bill and legs as adult. **Voice** Goose-like honks. **SS** Unmistakable. **Status** Historically, a rare visitor to mudflats of W Trinidad. However, since 2003 there have been several large groups in Jul–Aug. No records for Tobago.

Maguari Stork

juvenile

adult

Wood Stork

adult

Greater Flamingo

juvenile

adult

Jabiru

adult

immature

## NEW WORLD VULTURES – CATHARTIDAE

Seven species occur throughout the Americas. Just three species occur on Trinidad, one of which has been found on Tobago. Characterised by bare head and facial skin, long broad wings and effortless, endless soaring.

### Black Vulture *Coragyps atratus* 64cm; WS 145cm

Large, broad, slightly rounded wings with splayed 'fingers' and bulging secondaries. Tail very short and square. Deep rapid wingbeats; soars and glides on slightly raised or flat wings. Adult has dark grey, and very wrinkled, head and neck. Plumage black with most of outermost six primaries dirty white, visible on underwing. White feather quills visible on upperwing. Bill blackish; legs pale blue-grey. Immature recalls adult but duller, brownish-black. **Voice** Usually silent. **SS** From immature King Vulture by smaller size and white outer primaries. **Status** Abundant resident throughout Trinidad; gregarious, most common in open country, especially at dumps and fishing depots, where scavenges for food. Rare visitor to Tobago.

### Turkey Vulture *Cathartes aura* 69cm; WS 175cm

Long, broad parallel wings with splayed 'fingers'. Tail long and slightly rounded. Deep, slow wingbeats. Soars on pronounced dihedral, rocking from side to side. Adult has red bare head and facial skin. Underparts black, nape mottled creamy-yellow, rest of upperparts brownish-black. Underwing-coverts black contrasting with pale grey flight feathers. Bill pale grey; legs darker. Immature recalls adult but black head and neck skin. **Voice** Usually silent. **SS** From Zone-tailed Hawk by much larger size, greyish bill and legs, lack of barring on underwing flight feathers and lack of bands on tail. **Status** Common resident throughout Trinidad. Absent from Tobago.

### King Vulture *Sarcoramphus papa* 76cm; WS 190cm

Huge with broad, square-tipped wings, bulging secondaries and very short, square tail. Deep, strong wingbeats but usually soars on level wings with just tips upturned. Adult has colourful bare head and neck, a mixture of red, yellow and grey, with a grey neck-ruff and white eyes. Underparts white with black undertail. Mantle, rump and wing-coverts white; flight feathers and tail black. Bill black tipped red; legs whitish. Immature has dark grey or black head, neck and facial skin. Body plumage, wings and tail blackish, with dark bill and pale pinkish-grey legs. Gradually gains white, initially on underwing-coverts, but takes five years to attain full adult plumage. **Voice** Usually silent. **SS** Adult unmistakable. Immature from similar Black Vulture by larger size and lack of white on outer primaries. **Status** Very rare visitor to lowland forests of Trinidad, with just five records in last 12 years. No records from Tobago.

**Black Vulture**

**Turkey Vulture**

*ruficollis*

juvenile

adult

*jota*

*meridionalis* boreal migrant

**King Vulture**

adult

3rd year

2nd year

juvenile

adult

immature

## OSPREY – PANDIONIDAE

A monospecific family found in all non-polar continents. Plunge-dives for fish. One unconfirmed breeding report for Trinidad in 2006.

### Osprey *Pandion haliaetus*                                          57cm; WS 150cm

Narrow pointed wings angled at the carpal and short tail. Slow shallow wingbeats; soars on arched M-shaped wings. Hovers and dives into water to hunt. Adult at rest large but sleek with white crown and face stained brown on forehead, and broad dark brown stripe from ear-coverts to nape. Eyes yellow. Underparts white with variable loose gorget of brown upper-breast streaks. Upperparts chocolate-brown. Underwing-coverts white with black carpal bar and patch. Secondaries smudgy grey with dark barring; primaries cleaner with clearly defined barring; inner feathers translucent, outer ones tipped dark grey. Undertail densely barred grey. Cere and bill grey, tipped black; legs pale blue-grey. Immature has darker crown, buff wash to upper breast and underwing-coverts, and pale-scaled upperparts. **Voice** A series of short, loud whistles. **SS** Unmistakable. **Status** Common migrant to coasts and marshes of T&T in Sep–Apr; a few present year-round.

## HAWKS, KITES AND EAGLES – ACCIPITRIDAE

A large, diverse family of 240 species worldwide, variable in size and shape. A total of 24 species has been recorded in T&T. Of these, six have occurred on both islands in the last 10 years and 18 are found only on Trinidad. Of these, 17 are resident and the rest either migrants or vagrants. Identification requires an understanding of jizz and flight action, whilst underwing and undertail patterns are as important, if not more so, than body plumage.

### Hook-billed Kite *Chondrohierax uncinatus*                          44cm; WS 89cm

Broad, paddle-shaped wings pinched-in at base, and long tail. Slow wingbeats; soars and glides on flat or slightly bowed wings. Sluggish, secretive and shy, perching inside forest, but soars high over canopy. At rest appears small-headed, with large, hooked bill, long wings and tail. Polymorphic, but all share distinctive yellow loral skin. Adult male usually of grey morph, with pale grey crown and face, and white eyes; underparts barred grey and white; upperparts grey with white uppertail-coverts; and both upper- and undertail black with two broad pale grey bands and thin white tips. Underwing-coverts grey, flight feathers barred black and white. Adult female usually of rufous morph, with dark grey crown and face, rest of underparts barred rufous and white, nape rufous, and rest of upperparts dark brown with two pale grey tail-bands. Underwing-coverts barred rufous and white; flight feathers barred black and white. Undertail black with two broad pale grey bands. Dark morph rare: all black except broad white tail-band. Bill black, legs yellow. Most immatures have black crown with white face and underparts. Nuchal collar cream; rest of upperparts dark brownish-grey with rufous fringes. Tail black with several grey bands. Underwing recalls rufous morph female. Some immatures have similar head and tail pattern, but rest of body black with rufous fringes. **Voice** Usually silent; occasionally gives a musical whistle. **SS** From all similar hawks by yellow loral skin. Grey adult additionally from Grey-lined Hawk and Grey-headed Kite by bill shape and strongly barred underparts. **Status** Uncertain. Most records in T&T are in Aug–Apr and relate to visitors, but a very local resident population may persist in open lowland and secondary scrub in S Trinidad.

Osprey

Hook-billed Kite

adult
grey morph

adult
dark morph

juvenile

juvenile
dark morph

grey morph ♂

rufous morph ♀

adult
brown
morph

dark morph

immature ♂

immature ♀

### Grey-headed Kite Leptodon cayanensis cayanensis 49cm; WS 99cm

Broad, rounded wings with bulging secondaries and long tail. Distinctive flap-flap-glide flight action; glides and soars on flat wings. Adult at rest appears large, bulky but small-headed. Crown and face pale grey, underparts white, nape pale grey; mantle, rump and folded wing slate-grey, but upperwing has darker bands on flight feathers. Uppertail black with three narrow white bands; undertail white with same number of black bands. Underwing-coverts black; flight feathers banded grey and white. Cere grey, bill black, legs grey. Immature dimorphic. Dark morph has dark grey-brown crown and face; underparts creamy white variably streaked grey-brown. Upperparts dark grey-brown, some feathers fringed rufous. Both upper- and undertail broadly banded pale and dark grey. Underwing-coverts buff-white; flight feathers pale grey narrowly barred black. Pale morph has black crown. Nape, face and underparts white with thin black postocular stripe. Underwing and undertail paler. **Voice** Usually silent. **SS** Adult unmistakable. Pale morph immature from similar Hook-billed Kite by larger size, longer tail, smaller bill, absence of yellow cere, and lack of barring on underparts. **Status** Uncommon resident in forest at all elevations on Trinidad. Absent from Tobago.

### Swallow-tailed Kite Elanoides forficatus yetapa 61cm; WS 127cm

Long, narrow, angled, pointed wings and long, deeply forked tail. Glides and soars on flat wings. At rest, adult has entire head, nape and underparts white. Foremantle and upperwing-coverts black, rest of upperparts dark grey with variable white tertial fringes. Underwing-coverts white; flight feathers and undertail black. Bill blue-grey tipped black, legs blue-grey. Immature similar but shorter tailed with dark brown upperparts and streaked head. Boreal migrant race *forficatus* has purple gloss on back. **Voice** Usually silent; occasionally utters a high-pitched, shrill whistle. **SS** Unmistakable. **Status** Uncommon breeding visitor from mainland to Trinidad, in N and E forests between Feb and Aug. Just one recent record from Tobago, in Jun 2005.

### Pearl Kite Gampsonyx swainsonii leonae 22cm; WS 51cm

Narrow, pointed wings and rapid wingbeats; glides on flat wings. Adult at rest tiny and falcon-like with creamy-yellow forecrown; rest of crown grey. Face creamy yellow with black loral spot; underparts white, with chestnut flanks and thighs; nape white, with a broad chestnut nuchal collar, becoming black on neck-sides; and rest of upperparts slate-grey with random white feathers admixed, white wing-tips and trailing edge. Underwing-coverts white; flight feathers pale grey. Undertail white. Bill grey, legs yellow. Immature duller with browner upperparts fringed rufous; whiter hind-collar. **Voice** High-pitched, fast *kit kit kit* in alarm. **SS** Unmistakable. **Status** Uncommon resident of dry open savannas and suburban lowlands on Trinidad. Absent from Tobago.

### White-tailed Kite Elanus leucurus leucurus 40cm; WS 94cm

Long, pointed wings and squared tail. Deep elastic wingbeats; glides and soars on arched wings. Frequently hovers. Adult at rest appears small and dainty with grey-white crown. Face white with black around eyes. Underparts, including undertail white. Nape, mantle, rump and central tail pale grey; outer tail feathers white. Shoulder black; rest of upperwing pale grey. Underwing-coverts white with black carpal spot. Secondaries pale grey; primaries darker. Bill dark grey; legs yellow. Immature similar but has dark crown streaking and rufous tone to underparts; browner, mottled mantle and pale fringes to wing-coverts. **Voice** Usually silent; occasionally a soft whistle. **SS** Unmistakable. **Status** Accidental visitor to Trinidad with no documented records in last 10 years.

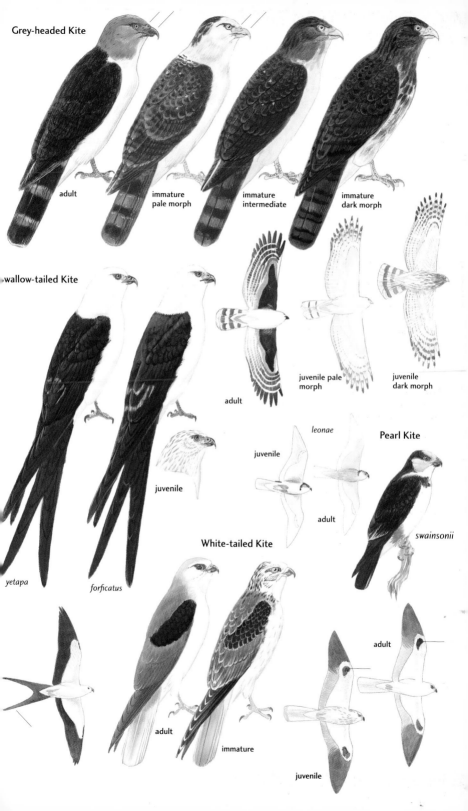

**Grey-headed Kite**

adult

immature pale morph

immature intermediate

immature dark morph

wallow-tailed Kite

adult

juvenile pale morph

juvenile dark morph

juvenile

*leonae*

juvenile

adult

Pearl Kite

*swainsonii*

*yetapa*

*forficatus*

White-tailed Kite

adult

adult

immature

juvenile

## Snail Kite *Rostrhamus sociabilis sociabilis* 44cm; WS 106cm

Long, broad rounded wings and short, slightly notched tail. Deep floppy wingbeats; glides on gently bowed wings. At rest appears medium-sized with wing-tips extending beyond tail. Adult male has sooty-black crown and face with red orbital and cere. Throat, breast and belly sooty-black, undertail-coverts and undertail white with broad black mid-tail-band. Nape, mantle, tail, upper- and underwing sooty-black contrasting with square white rump, uppertail-coverts and terminal tail-band. Bill sickle-shaped and black. Legs red. Adult female has chocolate-brown crown. Lores, forehead and face creamy white with reddish eyes and cere, and brown comma-shaped streak sweeping back from eye. Underparts creamy white, heavily blotched brown on breast and belly. Nape, mantle and wing-coverts chocolate-brown, coverts fringed and tipped buff. Rump white; tail brown, faintly tipped white. Upperwing blackish with white primary shafts. Underwing-coverts brown; flight feathers buff-white, heavily barred dark grey. Bill black, legs orange. Immature recalls adult female but has dark eyes and yellow legs. **Voice** A sheep-like bleating. **SS** Adult from melanistic Hook-billed Kite by undertail-coverts, rump, cere and leg colours. **Status** Rare visitor to freshwater marshes on Trinidad. No records for Tobago.

## Crane Hawk *Geranospiza caerulescens caerulescens* 47cm; WS 94cm

Rounded wings and long tail; flap-flap-glide action on flat wings recalls Grey-headed Kite. Adult at rest appears small-headed, small-billed, long-bodied and long-legged. Crown and face grey, slightly paler on forecrown, lower ear-coverts, chin and throat. Breast and belly grey; undertail-coverts pale grey-buff. Upperparts grey; flight feathers black. Both upper- and undertail black with two broad white bands and tip. In flight, underwing has prominent white crescent across mid-primaries. Bill and cere black; legs bright red. Immature has grey crown and face densely streaked white. Underparts pale buff, heavily streaked grey or brown. Upperparts as adult but with buff fringes to mantle and coverts. Legs duller orange. **Voice** A shrill whistle. **SS** Combination of white wing crescent, dark bill and reddish-orange legs unmistakable. **Status** Present on Trinidad since 2001, where now a very local rare resident in central savannas and breeding suspected. Absent from Tobago.

## Double-toothed Kite *Harpagus bidentatus bidentatus* 35cm; WS 66cm

Short, rounded wings and longish tail. Fast wingbeats; soars and glides on flat wings held slightly forward. Very accipiter-like. Adult at rest appears small and chunky with grey crown and face, and yellow orbital around orange eyes. Throat white with black mesial streak. Upper breast rufous-orange; lower breast and belly duller rufous, barred grey and white. Nape grey, rest of upperparts dark grey-brown with three white bars on uppertail. White undertail-coverts fluff out in flight and are prominent from above. Underwing white with black-barred primaries. Undertail black with three narrow white bars. Cere yellow-green, bill dark grey, legs yellow. Immature has brown crown and face streaked buff. Throat like adult, but rest of underparts cream with heavy brown streaking. Upperparts brown with pale fringes to mantle and coverts. Underwing and undertail as adult but with duller flight-feather barring. **Voice** A thin *tsip*. **SS** Unlikely to be confused once size and shape appreciated. **Status** Uncommon resident favouring forest at all elevations on Trinidad. Absent from Tobago.

juvenile

immature

Snail Kite

♂

Snail
Kite
adult

♀

♂

Crane
Hawk
adult

Crane Hawk

juvenile

juvenile

immature

adult

adult

Double-toothed Kite

juvenile

adult

adult

immature

immature

black morph

# PLATE 21: KITES IV AND HARRIERS

## Plumbeous Kite *Ictinia plumbea*                     35cm; WS 78cm

Long, pointed wings, slightly pinched-in at base and short, triangular tail. Deep elastic wingbeats; glides on flat wings and often-spread tail. Adult at rest appears small and sleek with wing-tips reaching well beyond tail. Crown and face pale grey with black eye surround. All underparts mid grey with paler throat. Nape mid-grey, mantle, wing-coverts and rump darker. Flight feathers blackish. Both upper- and undertail black with two narrow white bands. In flight, both upper- and underwing show conspicuous rufous panel in mid-primaries. Bill tiny and black, legs reddish-orange. Juvenile has pale grey crown and face heavily streaked black, blackish eye surround and short, pale supercilium. Underparts buff-white, streaked dark brown. Upperparts dark grey-brown fringed buff or white. In flight, inner primary patch is duller and paler, and tail has three narrow pale bars. Legs yellow. **Voice** A rich three-note whistle with emphasis on middle one: *zwee zeeee eu*. **SS** Unmistakable. **Status** Common widespread breeding visitor to Trinidad, favouring open and gallery forests, second growth and mangrove edges; often found on exposed perch. Most records Feb–Sep. Absent from Tobago.

## Long-winged Harrier *Circus buffoni*                     54cm; WS 137cm

Large yet light and buoyant. Long-winged and long-tailed. Slow elastic wingbeats; glides on raised wings; often hovers. Alternates long, low glides with drifting from side to side. At rest appears small-headed with long body, wings and legs. Plumage dimorphic. Pale morph adult male has black crown and face with white forehead, eyebrow, chin and lower cheeks. Throat white, bordered by broad black band bisected by thin white collar. Rest of underparts white with pale black flecks. Nape and mantle blackish with small, square white rump. Upperwing smaller coverts black; greater coverts and flight feathers silver-grey, barred and tipped black. Upper- and undertail banded pale grey and black. Underwing-coverts buff-white with faint dark barring; flight feathers silver-grey. Cere and bill grey, legs pale yellow. Adult female browner above and creamier below. Dark morph adult has head and underparts black with restricted white facial pattern. Upperparts, upper- and undertail as pale morph, but rump barred. Underwing-coverts black, flight feathers silver-grey, barred and tipped black. Immature pale morph recalls similar adult female but has rusty collar and pale fringes to wing-coverts. Immature dark morph recalls adult but dark grey underparts heavily streaked buff; mantle and upperwing-coverts fringed rufous, rump barred. **Voice** Usually silent. **SS** Unlikely to be confused. **Status** Locally uncommon resident of freshwater marshes, coastal mangroves, rice fields and open cultivation on Trinidad. Absent from Tobago.

Plumbeous Kite

adult

immature

juvenile

adult

juvenile

Long-winged Harrier

pale
morph

pale
morph

pale
morph

immature

pale
morph

♀

♂

immature

dark
morph

dark
morph

immature

♂

pale ♂

pale morph

pale morph

pale morph

♀

pale morph

immature

♂

immature

dark morph

adult retaining some
immature characters

## White Hawk *Leucopternis albicollis albicollis*　　　　51 cm; WS 107 cm

Soars on flat, rounded wings held forward, with bulging secondaries and very short tail. At rest, adult has white crown and face with contrasting black lores. Nape and underparts white. Mantle white, lightly spotted black. Rump white. Upperwing black with faint white covert tips and white 'landing lights'. Underwing-coverts white, flight feathers white, faintly barred, with black trailing edge and tips. Uppertail mainly black with broad white band at base and white tail tip. Undertail white with black subterminal band. Bill black; cere blue-grey; legs pale yellow. Immature similar but tinged buff, with black crown streaking. **Voice** A loud, harsh screech *kuh-eeeer*. **SS** Unlikely to be confused. **Status** Uncommon resident of forest and second growth at all elevations on Trinidad. Absent from Tobago.

## Grey-lined Hawk (Grey Hawk) *Asturina nitida nitida*　　　　40 cm; WS 84 cm

Short, rounded wings and relatively long tail. Rapid wingbeats; glides and soars on flat wings. Adult at rest small and compact with head and underparts pale grey, palest on crown, faintly but heavily barred darker. Upperparts unmarked grey. Uppertail black with three broad white bands. Underwing dirty white with faint grey barring on secondaries; undertail white with two grey bands and broad black subterminal band. Bill grey; cere yellow; legs yellow. Immature has buff crown, densely barred brown. Face buff-white with brown ear-coverts and malar stripe. Underparts buff-white with large dark brown tear-shaped blotches. Upperparts dark brown with buff or rufous covert fringes; narrow white uppertail-coverts patch and narrow black bars on uppertail. Upperwing has variable pale patch at base of primaries. Underwing-coverts buff; flight feathers grey-white with darker grey tips. Undertail white with numerous narrow grey bars. **Voice** High-pitched, drawn-out *wee-ouuuur*. **SS** Adult unmistakable. Immature from similar-age Broad-winged Hawk by face pattern, upperwing pale primary patch, white uppertail-coverts, paler primary tips on underwing and more clearly defined tail-bands. **Status** Common and widespread resident on Trinidad, usually below 500m. No documented records from Tobago in last 10 years.

## Rufous Crab Hawk *Buteogallus aequinoctialis*　　　　44 cm; WS 99 cm

Broad-winged and rather short-tailed. Slow wingbeats; glides and soars on flat wings. Adult at rest stocky with powerful shoulders. Head sooty-black with yellow cere reaching to eye and narrow yellow orbital. Rest of underparts dull rufous with faint black barring. Upperparts dark grey, fringed rufous. Folded flight feathers brighter rufous with thin black barring and prominent black trailing edge. Both upper- and undertail black with white central band and white tips. Underwing bright rufous with faint black coverts barring, bold black leading and trailing edges, and primary tips. Substantial bill yellow with black tip, legs yellow. Immature has brownish-grey crown; face creamy white, flecked black. Underparts cream boldly blotched dark brown. Upperparts dark grey variably fringed rufous or buff; buff panel on secondaries and black barring and narrow white terminal band on uppertail. Underwing-coverts cream, flecked dark brown. Flight feathers whitish, with indistinct dark grey tips, leading and trailing edges. Undertail dirty white with broad grey terminal band. **Voice** A loud melodious whistle. **SS** Adult unlikely to be confused. Immature from similar Savannah Hawk by habitat, smaller size, shorter wings and tail, paler underparts (especially flanks and thighs) and lack of distinct dark terminal tail-band. **Status** Rare visitor or perhaps resident in mangroves on coasts of W and SE Trinidad. Absent from Tobago.

adult

White Hawk

immature

adult

Grey-lined Hawk

immature

adult

juvenile

adult

Rufous Crab Hawk

juvenile

adult

adult

immature

## Great Black Hawk *Buteogallus urubitinga urubitinga* 61 cm; WS 124 cm

Long-necked with short broad wings, bulging secondaries and short tail. Slow, strong wingbeats; glides and soars on flat wings. Adult at rest large, powerful and noticeably long-legged. Body plumage including underwing black, except white uppertail-coverts, basal half and tip of both upper- and undertail. Cere dull yellow. Bill substantial and dark grey; legs yellow. Juvenile has pale buff head with dense grey streaking on crown and line of black streaks behind eyes. Underparts buff, blotched dark brown. Upperparts dark grey-brown with buff mantle and covert fringes, and black tail-bands. Underwing-coverts creamy white with irregular brown flecking; flight feathers white, barred pale grey, with dark grey primary tips. Undertail white with numerous very fine dark bars. **Voice** Prolonged high-pitched, shrill *weeeeuuur*. **SS** From Common Black Hawk, with care, by larger size, longer neck and legs, and shorter wings. Adult by white uppertail-coverts and extensive white in tail. Immature by plainer face and narrower, more numerous, bars on tail. Best separated by voice. **Status** Local and rare resident on Trinidad, favouring lowland forest and second growth. An uncommon resident in E Tobago.

## Common Black Hawk *Buteogallus anthracinus anthracinus* 56 cm; WS 117 cm

Slow, strong wingbeats; glides and soars on flat, rounded wings or held in very shallow V, with bulging secondaries and short, broad tail. Adult at rest appears bulky with entire plumage, including underwing, black except white mark at base of underside of primaries, broad central white tail-band (conspicuous in flight) and narrow white tail tip. Cere and substantial bill orange-yellow, the latter tipped black; legs yellow. Juvenile has dark brown crown streaked buff. Face creamy white with heavy dark brown eyeline and malar stripe. Underparts cream with tear-shaped brown blotches. Upperparts dark brown with buff fringes and obvious pale buff panel on inner primaries. Underwing has dark brown coverts and secondaries, buff inner primaries and black wing-tips. Upper- and undertail buff-white, distinctly barred brown, with broader terminal band. **Voice** High-pitched series of shrill *klee klee kleee kleee* notes. **SS** From Great Black Hawk, with care, by smaller stockier structure, shorter legs, longer wings, and shorter neck and tail. Adult by white mark on underside of primaries, lack of white uppertail-coverts and restricted white in tail. Immature by bolder face pattern, malar stripe and bolder undertail barring. Best separated by voice. **Status** Common and widespread resident on Trinidad. Very rare on Tobago; one documented record in last 10 years.

Great Black Hawk

adult

immature

juvenile

juvenile

adult

Common Black Hawk

adult

immature

juvenile

adult

## Savanna Hawk *Buteogallus meridionalis*      54cm; WS 129cm

Long, broad wings and shortish tail. Flaps with heavy wingbeats and glides on curved wings, and soars on flat wings. At rest has near vertical jizz and long legs. Adult has rufous crown and grey face. Underparts rufous, faintly and narrowly barred black. Nape and mantle grey with rufous fringes to feathers; rump dark grey. Upper- and undertail black with broad white central band and white tips. Shoulder rufous, wing-coverts blackish fringed rufous, tertials grey, flight feathers rufous tipped black. In flight upper- and underwings rufous with black trailing edge and wing-tips. Cere yellow, bill small and black, legs yellow. Immature has dark brown crown and cream face with buff supercilium. Underparts pale buff with coarse dark streaks, blackish blotches on flanks and rufous thigh feathering. Nape and mantle dark brown fringed buff; rump buff. Upperwing rufous, barred black, underwing pale and lightly barred, with dark grey trailing edge and wing-tips. Tail black with several white bars. **Voice** Usually silent; occasionally utters a shrill long scream. **SS** Adult unlikely to be confused. Immature with care from immature Rufous Crab-Hawk and Black-collared Hawk by lack of collar, more distinctly barred uppertail, distinct dark grey terminal undertail-band and long, pale yellow legs. **Status** Common, widespread resident open lowlands on Trinidad. Perches upright on poles, stumps, posts, fences, etc. for long periods. One undocumented sighting on Tobago.

## Black-collared Hawk *Busarellus nigricollis*      48cm; WS 129cm

Long, broad wings, bulging secondaries and short, fanned tail. Glides and soars on slightly bowed wings, and occasionally plunge-dives for prey in shallow water. Adult at rest appears small-headed and bulky-bodied. Head creamy white with sparse black flecking and prominent black upper-breast collar. Rest of underparts rufous-buff. Upperparts darker rufous with sparse black streaking. Upper- and undertail rufous, tipped white, with broad black subterminal band. Upper- and underwing have paler rufous-buff secondaries with black trailing edge, contrasting with black primaries. Bill and cere black; legs pale grey. Immature has whitish head with darker streaking than adult. Upper-breast collar broader and less well-defined. Rest of underparts buff-brown with dark barring on belly. Upperparts dark brown fringed buff; tail like adult. Underwing-coverts rufous-buff densely flecked black. Flight feathers buff-white with grey trailing edge and ill-defined dark grey wing-tips. Undertail buff-white with grey subterminal band and white tips. **Voice** Usually silent. **SS** Unlikely to be confused. **Status** Accidental to wetlands on Trinidad with just one documented record in last 10 years. No records for Tobago.

## Broad-winged Hawk *Buteo platypterus antillarum*      42cm; WS 84cm

Slightly round-winged with shortish tail. Deep wingbeats; glides and soars on flat wings. Adult at rest small and stocky with grey-brown head. Throat dirty white bordered by short black malar stripes. Breast and belly buff-white heavily barred rufous. Undertail-coverts white. Upperparts dark brown fringed buff; flight feathers darker, brownish-black. Upper- and undertail black with 2–3 broad white bands. Underwing-coverts pale buff with darker flecks; flight feathers white with pronounced black trailing edge, black wing-tips and faint grey secondaries barring. Bill grey, cere and legs yellow. Rare black morph unrecorded in T&T. Immature similar but has heavy dark brown underparts streaking, paler underwing lacking trailing edge and has grey undertail finely barred black. **Voice** Thin, high-pitched whistle. **SS** Adult unlikely to be confused. Immature from immature Grey-lined Hawk by darker underwing tips, lack of upperwing primary patch, white in uppertail-coverts and bold head pattern. **Status** Uncommon visitor to Trinidad, favouring Northern Range and NE forests, Oct–Apr. Numbers augmented by passage migrants. Perches lethargically in exposed position. Fairly common and widespread resident on Tobago, with strong passage noted Feb–Mar.

## Savanna Hawk

adult
note variation

adult

immature

juvenile

adult

## Black-collared Hawk

adult

juvenile

immature

juvenile

adult

## Broad-winged Hawk

immature

adult

black morph

adult

## Short-tailed Hawk *Buteo brachyurus brachyurus* 42cm; WS 94cm

Long, rounded wings and shortish tail. Stiff wingbeats; soars and glides on flat wings, occasionally held in a slight dihedral. Whilst hunting hangs in air, lowering in stages. Adult at rest appears small and compact. Dimorphic. Pale morph has black hood with white forehead. Underparts white. Upperparts dark grey with darker tail-bands. Underwing-coverts white; flight feathers pale grey, narrowly barred black with black wing-tips. Undertail pale grey, barred black. Dark morph has body plumage sooty-black except white forehead. Uppertail grey with several black bands. Underwing-coverts black; inner flight feathers grey barred black, with black trailing edge; outer feathers white, tipped black. Cere yellow; small bill basally yellow, distally black; legs yellow. Immature recalls respective adult but pale morph has pale fringes to upperparts and creamy tone to underparts, whilst dark morph has white-streaked face, lower breast and belly. **Voice** Usually silent; occasionally a piercing scream. **SS** Pale morph unmistakable. Dark morph from Zone-tailed Hawk by smaller size, shorter more rounded wings, more pronounced black trailing edge and less distinct tail-bands. **Status** Common and widespread Trinidad resident; pale morph much more numerous than dark. Just one record from Tobago, in 1964.

## Swainson's Hawk *Buteo swainsoni* 52cm; WS 127cm

Long, slender, pointed wings and long tail. Elastic wingbeats; glides and soars on wings raised in dihedral. At rest appears slender, long wings reach tail tip. Polymorphic. Pale morph adult has grey crown and face with white forehead and throat. Upper breast reddish-brown, lightly streaked; rest of underparts buff-white with faint flanks streaking. Mantle and wing-coverts dark grey-brown with darker flight feathers. Rump and uppertail-coverts whitish; tail grey, finely barred black with dark subterminal band. Underwing-coverts white; flight feathers grey with darker trailing edge and black wing-tips. Undertail pale grey, finely barred black. Cere and legs yellow, bill black. Rufous morph similar but has strong reddish barring on underparts and flecking on underwing-coverts. Dark morph sooty grey-brown, with obscurely barred under- and uppertail, pale throat and undertail-coverts. Underwing-coverts dark brown. Immature pale morph has buff-white crown and face heavily streaked black, black eyes and malar stripe. Underparts buff-white, with strong black blotches on upper breast and flanks. Upperparts dark brown fringed buff; tail grey, barred black. Underwing-coverts pale buff, spotted dark brown; flight feathers pale grey, barred dark, with black wing-tips and carpal patch. Rufous and dark morphs have darker, blotchier underparts and underwing-coverts. **Voice** Usually silent. **SS** Pale morph usually unmistakable. Dark morph from White-tailed Hawk by slighter jizz, different wing shape and lack of white breast markings. **Status** Very rare migrant to T&T, with four accepted records, in Mar–Jun 1998 and Oct 1999.

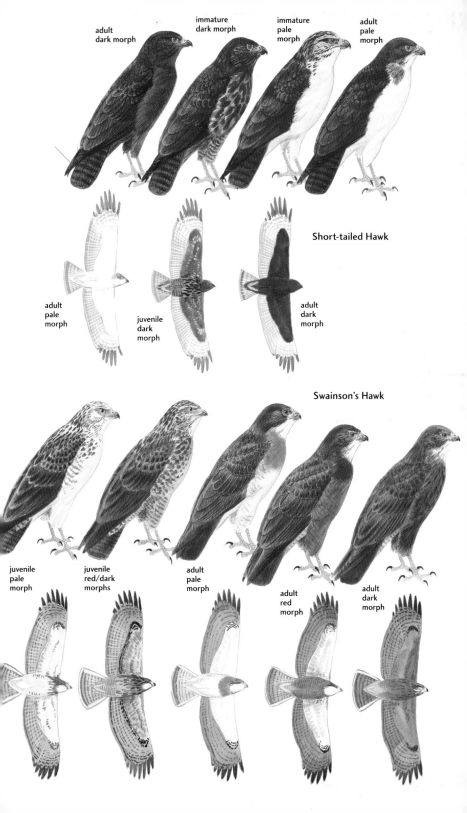

adult
dark morph

immature
dark morph

immature
pale
morph

adult
pale
morph

Short-tailed Hawk

adult
pale
morph

juvenile
dark
morph

adult
dark
morph

Swainson's Hawk

juvenile
pale
morph

juvenile
red/dark
morphs

adult
pale
morph

adult
red
morph

adult
dark
morph

## White-tailed Hawk *Buteo albicaudatus colonus*                          56cm; WS 129cm

Long wings slightly pinched-in at base, and short tail. Slow, deliberate wingbeats; glides and soars on raised wings. Often hovers. At rest, appears large and stocky, wing-tips project beyond tail. Dimorphic. Pale morph adult has grey crown and face with white forehead and chin. Throat variable, dark or white. Rest of underparts grey-white with variable grey flanks barring. Nape and mantle grey. Shoulder rufous, rest of upperwing grey with pale feather fringes and darker wing-tips. Rump, upper- and undertail white with broad black subterminal band. Underwing-coverts white with buff flecking; inner flight feathers grey, barred black with dark trailing edge. Outer feathers white, tipped black. Dark morph adult has small white forehead patch. Head and underparts darker grey with rufous-barred flanks and thighs. Upperparts dark grey with rufous shoulder. Tail as pale morph. Underwing-coverts mid-grey, variably flecked rufous. Bill black with yellow cere, legs yellowish-green. Immature has mainly sooty-black plumage, variable rufous fringes to upperparts and white patches on breast, undertail-coverts and uppertail-coverts. Underwing recalls dark morph adult; undertail white with numerous grey bars. **Voice** Usually silent. **SS** Adult unmistakable. Immature from dark morph Swainson's Hawk by underwing and undertail colour and markings, and white breast patch. **Status** Very rare visitor to open lowlands and freshwater marshes on Trinidad. No records for Tobago.

## Zone-tailed Hawk *Buteo albonotatus*                          51cm; WS 129cm

Long, parallel wings and long tail. Elastic wingbeats; glides with wings held in dihedral, strongly recalling Turkey Vulture. Adult at rest has sooty-black head, body and upperwing, occasionally with white blotches either side of rump. Uppertail black with several broad grey bands. Underwing-coverts black; flight feathers pale grey finely barred black, with thicker black trailing edge. Undertail black with broad white mid-band and narrower terminal band. Bill black, cere and legs yellow. Immature similar but brownish-black, with irregular white spots and paler undertail faintly but densely banded black. **Voice** A rarely heard high-pitched squeal. **SS** From Turkey Vulture by much smaller size, banded tail, barred underwing and conspicuously yellow bare parts. From dark morph Short-tailed Hawk by larger size, longer wings and more distinct undertail-bands. **Status** Common and widespread resident on Trinidad. One bird on E. Tobago, Dec 2006–Apr 2007.

## Black Hawk-Eagle *Spizaetus tyrannus serus*                          67cm; WS 132cm

Oval-shaped wings, distinctively pinched-in at base, and long tail. When gliding, wings held flat and slightly forward. Adult at rest appears large and powerful with flat crown and shaggy rear crest, tipped pale. Head sooty-black with small yellow eyes. Breast and belly sooty-black; tarsal feathers and undertail-coverts vermiculated black and white. Upperparts sooty-black. Underwing-coverts black with numerous small white spots; flight feathers extensively barred black and white. Upper- and undertail black with three pale bands. Bill grey; feet yellow. Immature has whitish head with dark ear-coverts. Upper breast white; rest of underparts buff heavily streaked black. Upperparts brownish-grey mottled black and white. Underwing-coverts white, heavily mottled black; flight feathers barred grey and white. Tail broadly tipped white. **Voice** Strident, clear 3–4-note *whee whee weee-er*. **SS** Size and distinctive flight silhouette render species almost unmistakable. **Status** Locally rare resident in lowland savanna and secondary forest on Trinidad; breeding suspected. Perches in subcanopy. Absent from Tobago.

White-tailed Hawk

juvenile

adult
pale
morph

adult dark
morph

adult
pale
morph

adult
dark morph

immature

adult

ult

juvenile

juvenile

juvenile

Zone-tailed
Hawk

adult

Black Hawk-Eagle

adult

juvenile

adult

immature

juvenile

### Ornate Hawk-Eagle *Spizaetus ornatus ornatus*       61 cm; WS 117 cm

In flight, protruding head and neck, short rounded wings pinched-in at base and long tail. Glides on level wings, held slightly forward. Adult at rest has black crown with shaggy erectile crest. Face, neck-sides and nape ginger. Throat and central upper breast white, bordered by broad black malar stripe. Rest of underparts, including feathered tarsi, white with broad black bars. Upperparts brownish-black with three broad grey tail-bands. Underwing white with black flecking on coverts; flight feathers white, tipped black and densely banded grey. Undertail black with three broad white bands. Cere yellow, heavy bill grey, feet yellow. First-year has white head and underparts with grey tips to crest feathers and occasional black flank bars. Upperparts brown with broad grey uppertail-bands and white tail tip. Underwing as adult, but lacks coverts markings. Face colour and underparts barring increase in subadult. **Voice** Far-carrying *whit wee eeeu wee wee wee*, with third syllable lower pitched. **SS** At rest unlikely to be confused. In flight, from Grey-headed Kite by much larger size and barred underparts. **Status** Locally rare resident in forested areas of Northern Range and NE Trinidad. No recent records for Tobago.

## CARACARAS AND FALCONS – FALCONIDAE

Sixty-four species worldwide, of which two caracaras and seven falcons recorded in T&T, but only two are resident; the rest migrants or vagrants. Many plumages are confusingly similar, therefore knowledge of size, shape and flight action is essential for correct identification.

### Yellow-headed Caracara *Milvago chimachima*       43 cm; WS 89 cm

Long, broad, round-tipped wings, and long tail. Flight, a series of shallow flaps followed by short glide. Adult at rest has creamy yellow head with black stripe from eye across ear-coverts, and bright yellow orbital and loral skin. Underparts creamy yellow. Upperparts chocolate-brown. Upper- and undertail cream with numerous narrow blackish bars and broad subterminal band. Underwing-coverts cream; flight feathers dark brown with two bold white patches at base of primaries. Small bill greyish, legs yellowish-green. Immature duller and browner with diffuse but heavy cream and brown streaks on head and underparts. **Voice** Long drawn-out scream with final notes lower. **SS** Unlikely to be confused. **Status** Common resident of lowland savannas, freshwater swamps and mangroves on Trinidad. Uncommon in SW Tobago.

### Northern Crested Caracara *Caracara cheriway*       56 cm; WS 119 cm

Long, broad, rather rectangular wings and long square-ended tail. Slow wingbeats; soars and glides on flat wings. Adult at rest appears large-headed with flat sooty-black crown and shaggy rear crest, long-necked and long-legged. Face, throat, neck and upper breast creamy white with blackish breast barring; lower breast and belly blackish. Undertail-coverts cream. Nape creamy-white; mantle, rump and wings black with pale mantle fringes. Upper- and undertail white with numerous narrow dark bars and broad black terminal band. Upper- and underwing dark brown with outer primaries white, heavily barred black. Bill and facial skin distinctive; bill distally pale blue, base, cere and lores pink. Legs pale yellowish-grey. Immature browner with less clean-cut pattern, also duller facial skin and fewer tail bars. **Voice** Usually silent, occasionally a harsh croak. **SS** Unlikely to be confused. **Status** Rare visitor to lowland Trinidad. No records from Tobago.

Ornate Hawk-Eagle

adult

juvenile

adult

juvenile

Yellow-headed Caracara

juvenile

adult

juvenile

adult

Northern Crested Caracara

uvenile

adult

juvenile

adult

## Aplomado Falcon *Falco femoralis femoralis*     41cm; WS 89cm

Medium size, long-winged and long-tailed, glides on flat wings, hunts very low over ground. Adult at rest has wings falling short of tail tip. Crown dark grey bordered by broad creamy white supercilium from eye and joining in V on nape. Face white with thick black moustachial, broad eye-stripe and yellow orbital. Throat and upper breast white tinged buff and streaked black, bordered by thick black mid-breast band that spreads along flanks. Belly, thigh feathers and undertail-coverts rich buff. Upperparts dark grey with heavy white tail-bands. Bill pale blue, tipped black, legs yellow. Upperwing dark grey with conspicuous white trailing edge. Underwing dark grey lightly but densely spotted white. Immature similar but duller, with supercilium, cheeks and upper breast buff, not white, and upperparts dark brown, faintly edged buff. **Voice** Usually silent. **SS** Unlikely to be confused. **Status** Rare visitor to Trinidad, favouring freshwater swamps and tidal mudflats where shorebirds roost. Most sightings Apr–Oct. Just one record from Tobago, Oct 2003.

## Bat Falcon *Falco rufigularis*     28cm; WS 58cm

Dashing and agile. Small with very quick wingbeats; long, thin, pointed wings and long tail. Adult at rest has complete sooty-black hood and broad yellow orbital. Chin white, becoming buff on throat and neck-sides. Breast and belly black, heavily barred white. Thigh feathers and undertail-coverts rufous. Upperparts sooty-black with indistinct white tail-bands. Wings sooty-black with heavy white underwing spotting. Cere yellow, bill blue-grey, legs yellow. Immature recalls duller adult, with black and buff underparts barring, and black-barred undertail-coverts. **Voice** Fast, excited scream: *kee kee kee.* **SS** From Orange-breasted Falcon by much smaller size and wing-tips fall well short of end of tail. **Status** Locally scarce resident to freshwater marshes, mangrove edges, savannas and secondary forest on Trinidad. Crepuscular; much of daytime spent perched in dead trees. Just one record from Tobago, Oct 2005.

## Orange-breasted Falcon *Falco deiroleucus*     36cm; WS 76cm

Medium size, powerful and thickset, with broad, pointed wings and rounded tail. Strong wingbeats; glides on flat wings. Adult at rest has wings projecting slightly beyond tail. Complete hood sooty-black, with yellow cere and orbital. Chin and throat white, neck-sides buff-white. Upper breast cinnamon, lower breast and belly black barred white or buff. Thighs and undertail-coverts cinnamon; undertail-coverts streaked black. Upperparts dark grey with blacker flight feathers and white tail tip. Underwing dark grey, densely spotted white. Bill grey, legs yellow. Immature has duller underparts and streaked upper breast. **Voice** A screaming *kyah kyah kyah.* **SS** From Bat Falcon by much larger size, larger bill, cinnamon upper breast and broader wings and tail. **Status** Accidental on Trinidad with no confirmed sightings in 30 years. No records from Tobago.

## Peregrine Falcon *Falco peregrinus*     43cm; WS 101cm

Large and powerful with broad, pointed wings, barrel-chest and short tail. Glides and soars on flat wings. Adult at rest has wing-tips almost reaching tail tip. Hood sooty-black with white lower cheeks, broad black moustachial, yellow cere and orbital. Throat and upper breast white becoming pale grey on lower breast, belly and undertail-coverts, with dense thick black barring. Upperparts, including upperwing, slate-grey. Underwing pale grey, heavily barred black. Bill grey, legs yellow. Immature similar, but browner above and streaked below. **Voice** Rough, hoarse, *kak kak.* **SS** From much smaller Merlin by much larger size, different face pattern and no wing barring. **Status** Fairly common visitor to both islands, found mainly in open lowland and suburban areas, Sep–Apr.

Aplomado Falcon

Orange-breasted Falcon

Bat Falcon

Peregrine Falcon

♂
immature

♂

immature

♂
immature

♀

♂
immature

tundrius

adult

cassini

immature

adult

anatum

adult

immature

immature

pealei

## American Kestrel *Falco sparverius isabellinus* 23cm; WS 56cm

Small, with long, pointed wings and long, rounded tail. Hunts with fast, shallow wingbeats; glides and soars on flat wings. Frequently hovers briefly. Considerable plumage variation. Adult male at rest has mainly blue-grey crown with inconspicuous reddish-brown central patch. Face white with broad black vertical malar and bar on rear ear-coverts, yellow orbital and cere. Throat white; breast and belly buff, variably spotted black. Flanks and undertail-coverts buff-white. Nape reddish-brown with large black spot on sides; mantle and rump reddish-brown lightly barred black. Upper- and undertail reddish-brown, with broad black terminal bar and white tip. Upperwing-coverts blue-grey spotted black; flight feathers blackish lightly spotted white. Underwing-coverts buff-white, spotted black; flight feathers white, barred black. Bill grey, legs yellow. Adult female has duller browner crown, creamy underparts streaked brown, and upperparts brown barred black, with less distinct subterminal tail-band and blackish flight feathers. Immature recalls adult. **Voice** A shrill *klee klee, klee*. **SS** Small size and head pattern renders it unlikely to be confused. **Status** Accidental visitor to Trinidad, with no records in last 10 years. Just one record for Tobago, Dec 1991.

## Common Kestrel *Falco tinnunculus* 31cm; WS 68cm

Medium sized, with long, narrow-based wings and long tail. Hunts with rapid shallow wingbeats; glides and soars on flat wings; frequently hovers. Adult male at rest has pale grey head with fine black crown streaking, black moustachial and thin yellow orbital. All underparts creamy-buff with bold black spotting on breast and upper belly. Nape, mantle and wing-coverts reddish-brown heavily spotted black. Flight feathers dark grey. Rump and upper- and undertail blue-grey with broad black terminal tail-band. Underwing-coverts pale buff spotted black; flight feathers pale grey, barred darker grey. Cere yellow, bill blue-grey with black tip, legs yellow. Adult female has brown crown, nape and mantle, less pronounced moustachial and black barring to grey-brown uppertail. Immature recalls adult female. **Voice** Series of short *kee kee kee* notes. **SS** From American Kestrel by larger size and different face markings. **Status** Accidental on Trinidad: immature female, Carli Bay, Dec 2003 (subspecies unknown).

## Merlin *Falco columbarius columbarius* 28cm; WS 63cm

Small, with short, broad-based, pointed wings and rather long tail. Flies, often very low, with rapid shallow wingbeats, very agile, twisting and turning when hunting. Glides and soars on flat wings. Adult male at rest has dark blue-grey crown. Face dirty grey with thin white supercilium, yellow cere and orbital ring. Throat and diffuse partial collar white, rest of underparts dirty white with dense reddish-brown streaks. Upperparts generally dark blue-grey with several black bars on uppertail. Underwing pale grey with densely spotted black coverts and dark grey-barred flight feathers. Bill small and black, legs yellow. Adult female has dark chocolate-brown crown; face paler with diffuse blackish malar stripe. Throat white, rest of underparts pale buff, densely streaked dark brown. Upperparts dark chocolate-brown. Underwing heavily spotted black and white; undertail black with narrow white bars. Immature recalls adult female. **Voice** High-pitched rapid *ki ki ki ki ki*. **SS** From Peregrine Falcon by smaller, slighter build, different crown and face markings, barred uppertail and flight action. **Status** Common visitor to open lowlands of T&T, Oct–early Apr. Race *richardsoni* paler, with more bars on tail, hypothetical for T&T, but confirmed visitor to mainland.

# American Kestrel

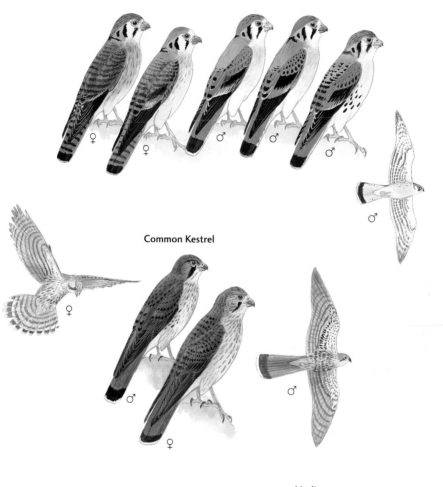

♀

♀

♂

♂

♂

♂

# Common Kestrel

♀

♂

♀

♂

# Merlin

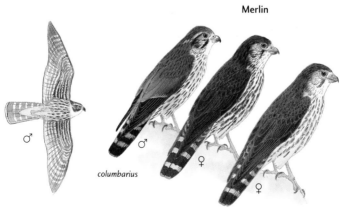

♂

♂

*columbarius*

♀

♀

*richardsoni*

# RAILS, CRAKES AND GALLINULES – RALLIDAE

Total of 136 species worldwide; 14 species recorded in T&T, of which 11 are resident, mainly on Trinidad. Characterised by short wings and tail, secretive and skulking habits. When alarmed, many quietly walk away through the undergrowth undetected, rather than flying. Rails are generally slim-bodied with rather long, slightly decurved bills and long legs. Wood rails are tall, long-legged and long-necked with slightly shorter, stouter bills. Crakes are small and dumpy with small heads and short bills. Most gallinules are larger and more colourful, and perch and swim in the open.

## Clapper Rail *Rallus longirostris pelodramus* 33cm

Adult has dark brown-black crown. Face grey marked with short white supercilium, black loral streak and reddish eyes. Throat white, upper breast dull reddish-buff. Lower breast and belly pale grey with diffuse dark grey and white flank barring. Undertail-coverts have white centres and dark grey sides. Upperparts warm brown, heavily streaked black; flight feathers dusky. Bill dull orange with dusky culmen and tip; legs pinkish-buff. Immature recalls adult but duller and plainer. Head grey with faint buff supercilium; bill grey. **Voice** Sharp series of notes like stones being rubbed together, *kyik kyik kyik*, often given in duet. **SS** Unlikely to be confused. **Status** Uncommon resident of mangroves, brackish and saltwater marshes, mainly on coasts of W Trinidad. Absent from Tobago.

## Grey-necked Wood Rail *Aramides cajanea cajanea* 36cm

Adult has grey crown and face with large reddish-brown eyes. Chin and throat white, upper breast and neck-sides grey. Lower breast and belly brick red, vent and undertail-coverts black. Rear crown blackish, neck and nape grey. Mantle, wing-coverts and inner flight feathers olive-brown, outer feathers rufous. Rump and tail black. Bill green with yellow base, legs coral-red. Immature recalls adult. **Voice** Loud, far-carrying *killik killok killik killok*. **SS** From Rufous-necked Wood Rail by crown, nape, underparts and leg colorations. **Status** Uncommon resident of mangroves, lowland wet forest, river edges and wet savannas on Trinidad; heard much more frequently than seen. Absent from Tobago.

## Rufous-necked Wood Rail *Aramides axillaries* 33cm

Adult has reddish-brown crown and face, with pale grey chin and red eyes. Throat grey, breast, belly and flanks reddish-brown, vent and undertail-coverts dark grey. Lower neck and nape grey, mantle and wing-coverts olive-green, flight feathers rufous. Rump and tail brown-black. Bill dirty apple green, legs dull pinkish-red. Immature has grey-brown head and underparts. **Voice** Recalls Grey-necked Wood Rail, a loud series of sharp notes, *qualp qualp qualp*. **SS** See Grey-necked Wood Rail. **Status** Rare resident in coastal mangroves of W Trinidad. Uncommon, but difficult to see in dry dense forest on Bocas Is. Absent from Tobago.

## Spotted Rail *Pardirallus maculatus maculatus* 28cm

Adult has black head, neck and breast densely spotted white, belly and undertail-coverts black, barred white. Upperparts dark brown streaked black and white. Bill greenish-yellow with red base to lower mandible, legs red. Immature browner with much-reduced underparts markings. **Voice** A four-note whistle and groaning screech. **SS** Unlikely to be confused. **Status** Very rare resident of freshwater marshes on Trinidad. Only two records in last 12 years, both roadside corpses. Its secretive habits may mask its true abundance. Absent from Tobago.

Clapper Rail

Grey-necked Wood Rail

adult

juvenile

Rufous-necked Wood Rail

adult

Spotted Rail

barred morph
juvenile

dark morph
juvenile

pale morph
juvenile

### Ash-throated Crake *Porzana albicollis olivacea*                                20cm

Olive-brown crown densely streaked black. Underparts mostly grey with paler, throat and rear flanks to undertail-coverts striped black and white. Upperparts black, broadly fringed olive-brown, appearing streaked. Bill dirty green, legs brownish. Immature recalls adult. **Voice** Likened to machine gun burst. **SS** From Paint-billed and Grey-breasted Crakes by larger size, upperparts streaking, bill and leg colour, and lack of chestnut hind-collar. **Status** Possibly a former resident of freshwater marshes on Trinidad. Lack of reports for at least 20 years suggests it is locally extirpated. Absent Tobago.

### Grey-breasted Crake *Laterallus exilis*                                14cm

Dark grey crown. Face, neck-sides and breast mid-grey; frontal whitish. Belly, rear flanks, and undertail-coverts barred black and white. Lower nape to upper mantle chestnut, rest of upperparts olive-brown with pale fringes on wing-coverts. Bill dull green, tipped black, legs greyish-green. Immature lacks chestnut nape. **Voice** Series of 5–6 piping notes like a dripping tap, *pee di di di di*. **SS** From Ash-throated and Paint-billed Crakes by small size and chestnut nape, and from Paint-billed also by bare-parts colour. **Status** Uncommon and extremely shy resident of freshwater wetlands on Trinidad. Disinclination to fly (sometimes even walking undetected between your feet!) may obscure true abundance. Absent Tobago.

### Paint-billed Crake *Neocrex erythrops olivascens*                                19cm

Adult has olive-brown crown, grey face, white throat and grey breast and belly. Rear flanks and undertail-coverts barred black and white. Upperparts olive-brown. Bill basally red, distally greenish-yellow; legs pinkish-red. Immature paler with greyish-yellow bill and greenish legs. **Voice** A high-pitched *keek* and frog-like *croak*. **SS** From Grey-breasted by larger size, lack of chestnut hind-collar and bare-parts colour. **Status** Very rare visitor to freshwater marshes on Trinidad. Just three records, the most recent Sep 2001, but secretive habits may mask true abundance. Absent Tobago.

### Sora *Porzana carolina*                                22cm

Breeding adult has black forecrown, foreface and throat; rear crown brown, rest of face grey with brownish wash to ear-coverts. Breast grey, paler on belly, becoming buff on undertail-coverts, outer feathers white. Flanks diffusely barred black and white. Brown above with black and white speckling on rump to tail. Bill yellow, legs yellowish-green. Non-breeder has restricted black on foreface and pale grey chin and throat. Immature has brown crown with narrow black central stripe. Face, neck-sides, throat and breast buff, belly and rear flanks patchily barred dark grey and cream. Upperparts similar to adult. **Voice** Usually silent in winter. Occasionally a long series of high-pitched thin notes, falling in pitch, *whee ee ee ee eeee e e e*. **SS** Unlikely to be confused. **Status** Scarce visitor to freshwater marshes in T&T, mostly Nov–Mar.

### Yellow-breasted Crake *Porzana flaviventer flaviventer*                                14cm

Black crown, face yellowish-buff with white supercilium and black eye-stripe. Throat white, breast and belly yellowish-buff. Lower underparts barred black and white. Upperparts brown-black, with sparse white and buff streaking and rufous tone to rump. Wing-coverts buff-brown barred black and white; flight feathers brown with buff fringes. Bill dirty yellow-grey, legs straw-yellow. **Voice** High-pitched and far-carrying *tee du*. **SS** Unlikely to be confused. **Status** Uncommon resident of freshwater marshes on Trinidad. Just one record, from Louis d'Or, NE Tobago, Dec 2002. Shy nature may mask true abundance.

## JACANAS – JACANIDAE

Characterised by long legs and toes, their feeding habits have earned them the name lily-trotters.

### Wattled Jacana *Jacana jacana jacana*                                24cm

Adult has black head and underparts. Upper mantle black, lower mantle, rump and tail warm cinnamon-brown. Upper- and underwing-coverts cinnamon with pale lemon-green flight feathers tipped black. Bill orange-yellow with red wattles on both sides extending onto forehead as a shield. Legs greenish-grey, dangling in flight. Immature lacks wattles, has a bold white eyebrow, and is entirely whitish below. Nape black; rest of upperparts bronze-brown. **Voice** Noisy, chattering rattle. **SS** Unmistakable. **Status** Abundant resident in lowland freshwater marshes on Trinidad; less common on Tobago.

## Ash-throated Crake

## Grey-breasted Crake

## Paint-billed Crake

adult

juvenile

## Sora

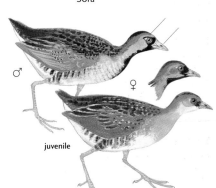

♂

♀

juvenile

## Wattled Jacana

juvenile

adult

## Yellow-breasted Crake

## RAILS, CRAKES AND GALLINULES (continued)

### Azure Gallinule *Porphyrio flavirostris* 25cm

Adult has dark grey-brown crown with apple green forehead shield. Face, neck-sides, breast and flanks aquamarine-blue, with reddish eyes. Throat and rest of underparts white. Mantle and wing feathers olive-brown with broad blue-grey fringes and distinct pale blue wash on shoulder. Rump and tail dark brown. Bill apple green, legs bright yellow. Immature has dark brown crown with dull green shield. Face, neck-sides and breast buff, rest of underparts white. Upperparts as adult. **Voice** Parrot-like *quawk* and various thin scratchy notes. **SS** Adult unmistakable. Immature with care from similar Purple Gallinule by smaller size, underparts colour and smaller, slighter bill. **Status** Scarce resident of Nariva swamp and Caroni rice project, Trinidad. Absent from Tobago.

### Purple Gallinule *Porphyrio martinica* 33cm

Adult has turquoise-blue shield. Rest of crown, head, breast and belly deep violet-blue with brighter blue neck-sides and flank-line. Undertail-coverts white. Upperparts bright greenish-blue. Bill thick and broad-based, bright red with yellow tip; legs bright yellow. Immature has dark brown crown with small, dull bluish-white shield, buff face and dirty white throat. Breast buff, belly, and undertail-coverts white. Rear neck buff, mantle, rump and tail olive-brown with dusky mottling on mantle. Wings dull bluish-green. Bill dull red, legs greenish-yellow. **Voice** Varied nasal cackles. **SS** Adult unmistakable. Immature from Azure Gallinule by much larger size, underparts colour and heavier bill. **Status** Common resident of freshwater marshes on Trinidad; uncommon on Tobago.

### Common Moorhen *Gallinula chloropus galeata* 36cm

Breeding adult has black crown and face, with bright red shield. Foreneck, breast and belly slate-grey, with prominent white flank stripe and white undertail-coverts with black central stripe. Hindneck black, mantle, rump and wing-coverts dark brown. Flight feathers and tail slate-grey. Bill red, tipped yellow, legs greenish-yellow. Non-breeder has duller shield, grey head and neck. Immature has brownish-grey crown and face with white chin and throat. Breast and belly drab grey with faint white flank stripe and white undertail-coverts with dark central stripe. Upperparts dull brownish-grey. Bill and legs greenish-grey. **Voice** Varied; utters clucks, screams and whines. **SS** Unlikely to be confused. **Status** Common resident of inland wetlands throughout T&T.

### Caribbean Coot *Fulica caribaea* 39cm

Adult has black crown and face with white bulbous frontal plate reaching well above eye level on forecrown. Upper breast black, lower breast and belly dark grey, undertail-coverts white. Upperparts dark grey with white trailing edge to inner flight feathers (noticeable in flight). Bill white with inconspicuous dusky subterminal band, legs greenish-yellow. Immature paler and duller. **Voice** Series of clucks. **SS** See American Coot. **Status** Very rare visitor to inland wetlands on both islands, with just six sightings in last 12 years; recorded Nov–May.

### American Coot *Fulica americana* 39cm

Adult has black crown and face with white forehead blaze reaching only to eye level. Bump of reddish facial skin on upper forehead. Breast, belly and flanks black, undertail-coverts white. Upperparts dark grey with white trailing edge to inner flight feathers (noticeable in flight). Bill white with black subterminal band, legs greenish-yellow. Eyes red. Immature slightly paler and duller. **Voice** Short clucking notes. **SS** Adult from Caribbean Coot with extreme care, by extent and shape of white frontal shield, and reddish forehead skin. Immature inseparable. Further caution necessary as the two species hybridise elsewhere in the West Indies. **Status** Very rare visitor to inland wetlands of SW Tobago, with just four sightings in last 12 years; recorded Sep–Feb. No records on Trinidad.

Azure Gallinule

Purple Gallinule

juvenile

adult

juvenile

adult

Common Moorhen

immature

Caribbean Coot

juvenile

adult

American Coot

## THICK-KNEES – BURHINIDAE

Characterised by large size, long legs, large head and short, straight bill. Nocturnal, with distinctive large eyes.

### Double-striped Thick-knee *Burhinus bistriatus vocifer*                46cm

Generally spends day lying on ground, becoming active and vocal at night. Grey crown bordered by broad black stripes. Face greyish-buff with bold white supercilium and large yellow eyes. Throat white, foreneck and breast greyish-buff with dense black streaks. Above brown, fringed buff. Flight feathers black with bold white band on inner primaries. Underwing white with broad dusky trailing edge. Tail dark brown, broadly fringed buff with subterminal white band. Bill yellow tipped black, legs yellowish-green. Immature paler. **Voice** At night a loud *kwee kwee*. **SS** Unmistakable. **Status** Very rare visitor to open savannas on both islands, with only five records in last 15 years; Aug–Feb, the latest being one on Tobago in February 2007.

## SUNGREBE – HELIORNITHIDAE

Sungrebes inhabit dark, shady waters in forests, and are secretive by nature.

### Sungrebe *Heliornis fulica*                30cm

Male has boldly patterned head and neck. Crown black, face white with thick black eyeline reaching to ear-coverts. Neck-sides and hindneck striped black and white. Foreneck, breast and belly white, with buff-brown wash to flanks. Upperparts olive-brown, tail blackish, tipped white. Bill dull red, black-and-yellow legs rarely seen. Female has buff lower face. **Voice** Usually silent. **SS** Unmistakable. **Status** Accidental to inland forested wetlands on Trinidad. Unrecorded since 1991 and never recorded on Tobago.

## OYSTERCATCHERS – HAEMATOPODIDAE

Boldly patterned, with long, straight, orange-red bills, they favour rocky and sandy beaches and estuaries.

### American Oystercatcher *Haematopus palliatus prattii*                45cm

Adult has black head and upper breast, with yellow eyes and narrow red orbital ring. Rest of underparts white. Mantle and rump brown. Inner wing-coverts brown; outer feathers white, forming bold stripe and contrasting with black flight feathers. Underwing white with dark outer flight-feather tips. Tail grey-white with broad black terminal bar. Bill orange-red, legs pale pink. Immature has duller red bill tipped black, greyer legs and pale fringes to upperparts. **Voice** A high, clear, piping *weep*. **SS** Unlikely to be confused. **Status** Very rare migrant to coastal mudflats of W Trinidad. Just 11 records in last 12 years, all Sep–Feb.

## STILTS AND AVOCETS – RECURVIROSTRIDAE

Large, graceful wading birds with extremely long legs. Stilts have long, straight, needle-thin bills and long wings projecting well beyond the tail, and avocets have long, upturned bills.

### American Avocet *Recurvirostra americana*                46cm

Breeding adult has brownish-orange head, neck and upper breast with paler lores. Rest of underparts white. Mantle, rump and tail white, contrasting with black inner scapulars. Wings boldly patterned black and white. Underwing-coverts and inner flight feathers white, outer feathers black. Bill black, legs blue-grey. Non-breeder has pale grey head, neck and upper breast. Immature like dull adult. **Voice** A sharp *kweep*. **SS** Unmistakable. **Status** Accidental on Tobago, with just one old and undated record.

### Black-necked Stilt *Himantopus mexicanus mexicanus*                36cm

Male has black head with conspicuous white patch around eyes; hindneck, mantle and wings black; lower mantle to tail and underparts white. Bill black, legs bright pink. Female has brownish wash to mantle. Immature has dark grey and white-mottled hindneck and greyish-brown mantle and tertials, with white fringes to feathers and white trailing edge to inner flight feathers. **Voice** Sharp, high-pitched *kik kik kik*. **SS** Unmistakable. **Status** Common breeding visitor to inland and coastal wetlands of Trinidad. Some year-round but most move to mainland Nov–Feb. Rare visitor to SW Tobago.

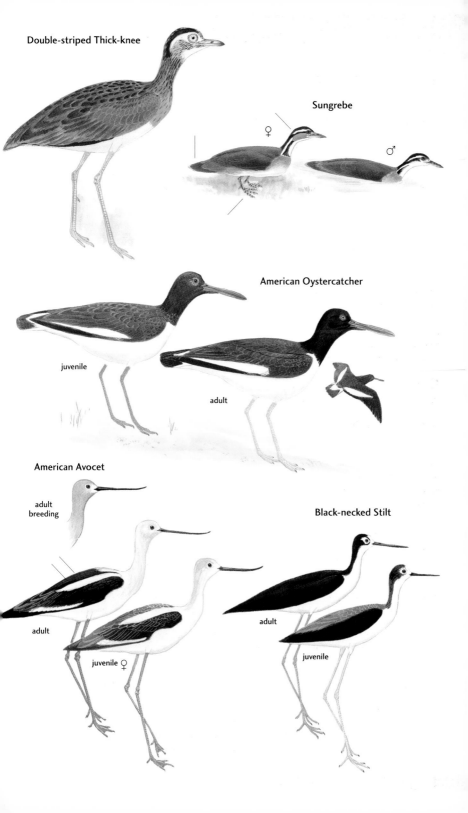

Double-striped Thick-knee

Sungrebe

♀

♂

American Oystercatcher

juvenile

adult

American Avocet

adult
breeding

Black-necked Stilt

adult

adult

juvenile ♀

adult

juvenile

## LAPWINGS AND PLOVERS – CHARADRIIDAE

A total of 66 species occurs worldwide. Plovers have a rather squat profile, with thick necks, long wings, short tails and short bills; lapwings are much more variable. Of the ten species that occur in T&T, just three are present year-round on both islands, with the rest either migrants or very rare visitors.

### Pied Lapwing *Vanellus cayanus* 23 cm

Small and dainty. Adult has pale grey-brown crown bordered by broad white stripe meeting on nape. Forehead, nape and face black with red orbital ring. Underparts white with broad black pectoral band. Mantle pale grey-brown with broad black scapular lines. Inner wing-coverts pale grey-brown, outer feathers and inner flight feathers white contrasting with black outer flight feathers. Underwing-coverts and inner flight feathers white, rest black. Rump and basal half of tail white, distal half black. Bill black, legs coral-red. Juvenile has white forehead, dark grey breast-band, brown scapular lines and extensive pale fringes to mantle and coverts. **Voice** Two-note plaintive *chee wee*, the second note pitched lower. **SS** Unlikely to be confused. **Status** Accidental on Trinidad, with just one record, in Mar 1998.

### Southern Lapwing *Vanellus chilensis cayennensis* 35 cm

Large and long-legged with broad, rounded wings. Light, springy wingbeats, but slow-flapping display-flight. Grey-brown crown, with short, thin, trailing black crest. Forehead and chin black, narrowly bordered white, face grey-brown with red eyes. Throat and upper breast black with grey-brown neck-sides; rest of underparts white. Nape and mantle grey-brown with bronzy shoulders. Rump white, tail black, tipped white. Bill pink tipped black, legs dark reddish-brown. Juvenile has buffy crown speckling, lacks white on foreface and bronze shoulders, has a grey throat and upperparts fringed buff. **Voice** Very noisy; a high-pitched, penetrating *kee-ow* uttered day and night. **SS** Unmistakable. **Status** Abundant, widespread resident of lowland grasslands and wetlands on Trinidad; less common on Tobago.

### American Golden Plover *Pluvialis dominica* 27 cm

Small head and short, thin bill. Wings project well beyond tail. In flight shows wing-stripe but no rump patch. Breeding adult has black crown feathers broadly fringed golden-green. Face black. Broad white forehead extends as border to crown and face and down neck- and breast-sides. Rest of underparts black. Upperparts black, fringed golden-green or white. Flight feathers dark grey with faint white wing-stripe and shafts to outer primaries. Underwing smoky grey. Bill black, legs dark grey. Non-breeding adult has brownish-grey head with broad white supercilium behind eye. Underparts pale grey variably blotched brown or black. Upperparts dark grey, narrowly fringed buff or white. Upper- and underwing as breeders. Immature paler with pale grey crown densely streaked black. Face grey-buff with strong white supercilium and large dark eyes. Underparts grey-white with diffuse buff bars and chevrons. Nape and mantle pale grey-buff, streaked darker. Scapulars, wing-coverts and rump black, broadly fringed pale buff. **Voice** A clearly disyllabic *clu-ee*. **SS** At all ages, from Black-bellied Plover by smaller size and shape, bill- and wing-lengths, underwing colour, lack of distinctive rump patch and axillaries. **Status** Uncommon passage migrant to wet grasslands and wetlands across Trinidad, mostly Jul–Oct. Decidedly scarcer on Tobago.

### Black-bellied (Grey) Plover *Pluvialis squatarola* 29 cm

Chunky with large head and stout bill; wings reach tail tip. In flight has both wing-stripe and rump patch. Breeding adult has white crown and upper face to eye level, with faint grey crown flecking. White extends down neck-side, onto breast-sides. Lower face, throat, central breast, belly and flanks black, and undertail-coverts white. Mantle and upperwing-coverts spangled grey, black and white, upperside of flight feathers dark grey with white wing-stripe. Underwing white with black axillaries, dusky trailing edge and wing-tips. Rump and tail white with narrow grey tail barring. Bill black, legs grey. Non-breeding adult has grey crown densely mottled black. Face pale grey with diffuse white supercilium and lores, and dusky ear-coverts. Throat white, breast grey lightly streaked darker. Belly and undertail-coverts grey-white. Mantle grey, lightly barred black, wing-coverts spangled grey, black and white. Rump white, tail barred grey and black. Immature recalls non-breeding adult with neat white fringes to upperparts. **Voice** Mournful whistle, *pee u eee*. **SS** At all ages, from American Golden Plover by larger overall size, bill size, wing-length, undertail-coverts and rump colour, upperwing-stripe and black axillaries. **Status** Common non-breeding visitor, primarily to tidal mudflats of W Trinidad, rarely at inland wetlands. Present year-round but mostly Aug–Oct, Apr–May. Decidedly uncommon on Tobago.

Pied Lapwing

juvenile

adult

Southern Lapwing

juvenile

adult

American Golden Plover

winter

mid-moult

breeding

Black-bellied Plover

juvenile

breeding

non-breeding

### Common Ringed Plover *Charadrius hiaticula tundrae* 18cm
Small and dumpy with wings more or less reaching tail tip. In flight, wing-stripe but no rump patch. Breeding adult has white forehead and black band over crown, through eyes to ear-coverts. Upperparts brown, except narrow white collar. Broad black breast-band, rest of underparts white. Non-breeding adult has pale cinnamon forehead and reduced breast-band, juvenile has partial brown breast-band. **Voice** A rising *poo-eep*. **SS** Separable with extreme care from Semipalmated Plover by larger size, wing-length, bill shape and size, and lack of distinct orbital ring and semipalmations. Breeding adult by distinct supercilium and broader breast-band. Immature by close examination of feathering around gape. Easiest by voice. **Status** Accidental W Trinidad; one trapped, Oct 1962.

### Semipalmated Plover *Charadrius semipalmatus* 18cm
Very similar to slightly larger Common Ringed, the key difference being much shorter post-ocular line, but bill smaller and breast-band narrower in all plumages. Wings more or less reach tail tip, Legs orange-yellow, with semipalmations between central and inner toes. Non-breeding adult has duller brown crown and face markings, diffuse whitish supercilium extending behind and below eye, browner breast-band and mainly black bill. Immature is paler above with paler coverts fringes, and yellower legs. **Voice** Single *chew-üt*. **SS** From Common Ringed by longer wings, shorter, stubbier bill, narrow orbital ring, and semipalmations. Breeding adult lacks obvious white supercilium and has narrower breast-band. Immature and non-breeder by loral pattern and supercilium shape. Best separated by voice. **Status** Common non-breeding visitor to inland wetlands and coastal mudflats of Trinidad. Present year-round, with most Aug–Apr. Much less common on Tobago.

### Collared Plover *Charadrius collaris collaris* 14cm
Small and slim; very clean-cut with thin bill. Wing-tips fall just short of tail. In flight, wing-stripe but no rump patch. Typical *Charadrius* plover, but separated from others by lack of white nuchal collar. Male has forehead white, crown black and rear crown cinnamon, whereas female has duller cinnamon on head, immature lacks cinnamon, and juvenile has unpatterned head. **Voice** Rather soft *teet* or *chit*. **SS** From other ringed plovers by small size, lack of white collar, face markings and upperparts colour. **Status** Common breeding visitor to coastal and inland wetlands of Trinidad, much scarcer on Tobago. Mostly late Apr–late Oct.

### Killdeer *Charadrius vociferus vociferus* 27cm
Attenuated jizz, with steep forehead, wedge-shaped tail and long legs. Feeds with tip-up action. Wings fall short of tail tip. In flight, wing-stripe but no rump patch. Largest *Charadrius* and unique in having two breast-bands with striking orange rump and uppertail-coverts, whilst non-breeder has rufous fringes to upperparts. Immature has dusky breast-bands and pale fringes to upperparts. **Voice** Loud strident *tee eeee* or *kee dee*. **SS** Unmistakable. **Status** Very rare visitor to open savannas and coastal and inland wetlands in T&T, with only three records in last 12 years; Jan–Mar.

### Wilson's Plover *Charadrius wilsonia cinnamominus* 19cm
Bulky with large, square head, thick neck and thick, bulbous bill. Wings reach tail tip. In flight, wing-stripe but no white rump patch. Feeds slowly and deliberately. Two distinct races, resident *cinnamominus*, with cinnamon face-sides and breast-band, and boreal migrant *wilsonia*, with dull brown head and breast-band (female) or black breast-band (male). White nuchal collar broken on nape, rump-sides and outer base of tail white. **Voice** Sharp *weep*. **SS** From other ringed plovers by robust jizz, bill shape and leg colour. **Status** Uncommon visitor to coasts of W Trinidad, mostly Jul–Oct. Has bred. Just one record Tobago, Apr 1996.

### Snowy Plover *Charadrius alexandrinus nivosus* 14cm
Small and pale with short wings just reaching tail tip and short thin bill. In flight, narrow wing-stripe but no rump patch. Distinctive, faster feeding action than other plovers. Soft grey above, all white below, with broad white hind-collar. Breeding male has black forecrown and ear-coverts. **Voice** A low-pitched *pi pee pi*. **SS** From immature Collared Plover by white nuchal collar, thinner bill and black legs. **Status** Accidental NE Tobago, Aug 1968.

**Common Ringed Plover**

juvenile

non-breeding

breeding

**Semipalmated Plover**

juvenile

non-breeding

breeding

**Collared Plover**

juvenile

immature

♂

**Killdeer**

♀

*cinnamominus*
(N S Am resident)

♂
(May)

**Snowy Plover**

juvenile

♂
breeding

**Wilson's Plover**

*wilsonia*
(Boreal migrant)

juvenile

♂

*wilsonia*

♀

## SANDPIPERS AND ALLIES – SCOLOPACIDAE

Small to medium-sized mud-loving waders, with short to long legs; the bill is often long, but comes in a highly diverse range of shapes and sizes.

### Wilson's Snipe *Gallinago delicata* 28cm

Dumpy, pot-bellied and short-legged. Large-headed and large-eyed, with long, straight bill usually held downwards. Wings fall short of tail tip. Usually freezes when disturbed, but may dash for cover. Flushes explosively at last minute in zigzag flight. Typical snipe with black-streaked chestnut upperparts, whitish below with buffy and black breast streaks and bars on flanks and undertail-coverts. Note pale-edged black tramlines on back, narrowly barred secondaries and two black bars on tail. **Voice** A long, nasal *skaaah* when flushed. **SS** Extremely hard to separate from next species, but has shorter bill and longer wings with no black subterminal tail tip. Only snipe on Tobago. **Status** Non-breeding visitor to agricultural land and freshwater marshes on both islands, Aug–Apr.

### South American Snipe *Gallinago paraguaiae paraguaiae* 28cm

Almost identical at all ages to Wilson's Snipe. In the hand most are distinguishable by width of axillaries and tertial bars, greyish tone to buff uppertail-coverts, and bill- and wing-length. **Voice** A harsh, rasping *kaatch* on flushing. **SS** Very hard to separate from Wilson's, but has longer bill, shorter wings, more obvious white wing-bar, a whiter supercilium and less brownish underwing. **Status** Local resident of lowland agricultural land, savannas and inland freshwater marshes on Trinidad. Very reluctant to flush. Absent Tobago.

### Short-billed Dowitcher *Limnodromus griseus griseus* 28cm

Stocky and short-legged with long, straight bill. Wings reach tail tip. In flight, lower mantle and rump wedge, contrasting pale inner flight feathers but no wing-stripe. Forages by walking slowly in mud, with distinctive sewing-machine feeding action. Non-breeding adult brownish-grey above with pale grey fringes to wing feathers, diffuse supercilium reaching to rear ear-coverts, and white below, washed grey on breast with pale, irregular barring on flanks and undertail-coverts. Lower back white, spotted black, broadening on rump and heavily barred black. Legs green. Breeder much darker and browner, and more boldly marked below. Juvenile brighter than breeding adult. **Voice** A clear, mellow *tu tu tu*. **SS** Unlikely to be confused. Note, Long-billed Dowitcher *L. scolopaceus*, a vagrant to mainland coasts and offshore islands, but not T&T, is best distinguished by call. **Status** Common non-breeding visitor to wetlands on both islands. Some present year-round, but mostly Aug–Apr.

### Hudsonian Godwit *Limosa haemastica* 39cm

Long legs, long, slender pointed wings that project slightly beyond tail and long, slightly upturned bill. In flight, wing-stripe and rump patch. Non-breeding adult has generally grey upperparts and head to breast, and white underparts, forehead and supercilium. Underwing-coverts black and tail black with white tip. Long, slightly upturned bill pink at base, with long, dark grey legs. Juvenile buffy, browner and spotted and barred on upperparts. **Voice** Usually silent. **SS** From Black-tailed by upturned bill, black underwing and white tail tip, and from Willet by longer, upturned bill and darker legs. **Status** Uncommon passage migrant to wetlands on Trinidad, mostly Aug–Oct. No recent sightings on Tobago.

### Black-tailed Godwit *Limosa limosa limosa* 42cm

Large with long legs, broad wings that extend slightly beyond tail tip and long, straight bill. In flight, wing-stripe and rump patch. Very similar to Hudsonian Godwit but centre of forehead grey or brown, bill straight, tail entirely black, upperwing-stripe broader, and underwing white. **Voice** A quick, nasal *vi vi ve*. **Status** Accidental W Trinidad, Sep 2000.

### Marbled Godwit *Limosa fedoa fedoa* 46cm

Large and bulky with long legs, broad wings that project slightly beyond tail and long, slightly upturned bill. Entirely warm ochre, with short white eyebrow, brown-streaked head and neck, becoming arrowheads on breast; rump and uppertail-coverts white and tail barred black. Distinct blackish patch on primary coverts. Long bill basally deep pink and long grey legs. Juvenile similar, but adult more boldly marked below. **Voice** A harsh *kra-wep*. **SS** Unlikely to be confused. **Status** Rare passage migrant to coastal mudflats of W Trinidad, with just five documented records in last 12 years. Most late Sept–Apr, once Aug (2006). No authenticated records from Tobago.

Wilson's Snipe

South American Snipe

Black-tailed Godwit

juvenile

Short-billed Dowitcher

non-breeding

juvenile

non-breeding

Hudsonian Godwit

Marbled Godwit

juvenile

non-breeding

## Whimbrel *Numenius phaeopus hudsonicus* 45cm

Large with long decurved bill. Adult has creamy central crown-stripe, with dark brown lateral coronal bands. Face pale grey-buff with creamy supercilium and dark brown eye-stripe. Underparts buff-white with dark brown neck and breast streaking and flanks barring. Underwing-coverts buff, barred black; flight feathers darker grey. Bill black, lower mandible base pinkish. Legs blue-grey. Juvenile brighter and neater. European race *N. p. phaeopus*, a very rare visitor to T&T, is paler and greyer, with white rump extending as wedge onto back, and has white underwing-coverts. **Voice** Rapid fluty *pi-pi-pi-pi-pi-pi-pi* on even pitch. **SS** From smaller Eskimo Curlew by more distinct crown and face markings, underparts and underwing colour. **Status** Common non-breeding visitor to coastal mudflats of Trinidad, but rare inland. Decidedly less common on Tobago. Present year-round, but most Aug–Nov.

## Long-billed Curlew *Numenius americanus* 58cm

Very large with broad wings and very long, decurved, thin bill. Wings reach tail tip. Brown crown finely streaked black. Face pale grey, finely streaked black with brown eye-stripe. Underparts cinnamon-buff faintly streaked brown on breast and fore flanks. Underwing cinnamon-buff with faint black flight-feather barring. Bill black with reddish base to lower mandible, noticeably longer and straighter-looking in female. Legs bluish-grey. Juvenile has shorter bill and lacks underparts streaking. **Voice** A short rising whistle, *coo-li*, with emphasis on second syllable. **SS** Unmistakable. **Status** Accidental on Tobago; two at Buccoo, Sep 1988 (subspecies unknown).

## Upland Sandpiper *Bartramia longicauda* 31cm

Attenuated long-legged profile and erect stance. Head small and rounded, bill short and straight, neck long and thin. Wing-tips fall well short of tail tip. Buff crown densely streaked black with faint pale central stripe. Face buff and rather plain with paler lores and lower cheeks, and narrow white orbital around large black eyes gives startled expression. Throat whitish, neck buffy and finely streaked, becoming more densely so on hindneck; back and wing-coverts dark brown, fringed golden-buff. Underwing heavily barred brown and buff. Rump and tail dark brown with white sides. Bill yellow with black tip, legs straw-yellow. Juvenile similar, but has indistinct flank markings. **Voice** Alarm-call in flight, a faint *quip pi de de*. **SS** Unmistakable. **Status** Very rare visitor to grasslands and rice fields in T&T, with seven records in last 12 years; Apr, Sep–Oct.

## Eskimo Curlew *Numenius borealis* 36cm

Small curlew with short, decurved bill. Wings extend well beyond tail. Narrow creamy central crown-stripe with diffuse dark brown lateral coronal bands. Face grey-buff with obscure brown stripe behind eye and dark lores. Throat white, neck and breast ochraceous with fine dark brown streaks, back and wings to tail dark brown, fringed and barred ochre; belly to undertail-coverts white. Underwing-coverts rich cinnamon-buff, spotted black; flight feathers dark grey. Bill black, with yellow base to lower mandible, legs pale blue-grey. Juvenile cleaner and neater. **Voice** Series of twittering notes. **SS** From Whimbrel by much smaller size, less distinct head and face markings, underparts and underwing colour. **Status** Critically Endangered, with no authenticated records since 1964. Historically, a passage migrant through T&T, but no reports there since late 19th century.

## Ruddy Turnstone *Arenaria interpres morinella* 24cm

Stocky and short-legged with short, sharp, straight bill. Forages on rocks, shingle and mud, hurriedly tossing seaweed and pebbles aside in search of prey. Very variable plumage, with various intermediates between full breeding and non-breeding. Black of breast always cleanly separated from pure white underparts. Bill black, legs orange. Very distinct pattern in flight, with white lower back to tail-coverts, black band across rump and broad subterminal black band on white tail. Female duller. Non-breeders lack chestnut in wings and black and white on face. Juvenile like non-breeder but appears scaly with neat white or buff fringes to mantle and wing-coverts. **Voice** Low-pitched chuckle and faster chattering rattle. **SS** Unmistakable. **Status** Non-breeding visitor to coasts of T&T. Whilst some remain year-round, most records Sep–Apr.

Whimbrel

Long-billed Curlew

Upland Sandpiper

Ruddy Turnstone

juvenile

non-breeding

breeding

Eskimo Curlew

### Greater Yellowlegs *Tringa melanoleuca* 36cm

Tall and thin-necked with long, slightly upturned and thick-based bill substantially longer than head. Wings project just beyond tail. In flight, rump patch but no wing-stripe. Breeding adult has white head, neck and breast, heavily streaked dark grey, with pronounced white orbital and faint white supercilium. Rest of underparts white with bold black flanks barring. Mantle, wing-coverts and inner flight feathers dark brownish-grey, spotted white, rest blackish. Underwing grey-white and faintly barred, darker at tip. Rump and tail white with dark central tail barring. Bill basally grey, distally black. Legs bright yellow. Non-breeder has greyer head with broader, more diffuse white supercilium and dark lores. Neck and breast densely streaked grey, rest of underparts white. Mantle and wing-coverts grey, spotted white. Immature similar but has more defined breast streaking and clean buff fringes to scapulars and wing-coverts. **Voice** Far-carrying *tieu tieu tieu*, much like Common Greenshank. **SS** From Common Greenshank by leg colour and lack of white mantle wedge. From Lesser Yellowlegs in comparative views by larger size. **Status** Common non-breeding visitor to wetlands across T&T, with some present year-round, but most numerous Aug–Oct, Mar–Apr.

### Lesser Yellowlegs *Tringa flavipes* 27cm

Medium size, slender and rather dainty with relatively short, thin straight bill. Wings project well beyond tail. In flight, rump patch but no wing-stripe. Plumage at all ages virtually identical to Greater Yellowlegs but has slightly more obvious white supercilium, plain blackish inner flight feathers and all-black bill. **Voice** Recalls Greater Yellowlegs but usually only 1–2 softer whistles, *tew tew*. **SS** From Greater Yellowlegs by smaller, daintier profile, bill shape, length and colour, wing-length and colour of inner flight feathers. **Status** Very common non-breeding visitor to wetlands across Trinidad, where present year-round but abundant Jul–Oct. Common on Tobago.

### Spotted Redshank *Tringa erythropus* 31cm

Rather large but slim and elegant with long, straight bill and long legs. Wings reach tail tip. In flight, lower mantle wedge, but no wing-stripe. Non-breeding adult has grey crown, pale grey face with prominent white supercilium and blackish lores. Underparts white with pale grey wash to throat and upper breast, and diffuse grey rear-flanks barring. Upperparts grey with narrow white rump and lower mantle wedge. Upperwing-coverts grey, fringed white; flight feathers darker. Bill dark with red base, legs bright red. Breeder has sooty-black head and underparts. Immature is darker grey with extensive white barring and yellowish-orange legs. **Voice** Loud *chew-it*, recalling Semipalmated Plover. **SS** Unlikely to be confused. **Status** Accidental Tobago, Feb 1983.

### Common Greenshank *Tringa nebularia* 33cm

Large with long legs, rather horizontal gait and long upturned bill. Wings reach tail tip. In flight, lower mantle and rump wedge but no wing-stripe. Breeder has head, neck to upper breast and flanks white, well streaked blackish. Above greyish-brown with black scapulars dotted white on their fringes, and some other scattered black feathers similarly marked; lower back to uppertail-coverts forms long white wedge that is very noticeable in flight; tail white with grey barring on central feathers. Bill grey-green, darkening at tip, legs dull green or greenish-yellow. Non-breeder appears paler and greyer, lacking any brown or black feathers on upperparts, and has reduced underparts streaking. Legs grey-green. Immature recalls non-breeder with neat white upperparts fringes. **Voice** Distinctive, loud *chew chew chew*. **SS** From Greater Yellowlegs by grey-green legs and white on lower back. **Status** Two undocumented but multi-observer sightings from SW Tobago, Jul 1977, and C Trinidad, early 1987.

### Red Knot *Calidris canutus rufus* 27cm

Stocky, short-legged and pot-bellied, with short, thick, straight bill and long wings. In flight, wing-stripe but no rump patch. Non-breeder has grey head with thin white supercilium and black eye-stripe. Breast grey with dark mottling, rest of underparts dirty white with grey flanks barring. Upperparts plain grey. First-year quite distinct, being flushed pale cinnamon from head to breast, streaked lightly with brown from crown to hindneck, and scalloped white on breast. Above dark brown with grey and cinnamon fringes. Breeder richer cinnamon. **Voice** Usually silent. **SS** Unlikely to be confused. **Status** Uncommon visitor to wetlands across Trinidad; some overwinter. Most records Aug–Oct, Apr–May. Decidedly scarce on Tobago.

**Greater Yellowlegs**

non-breeding

breeding

**Lesser Yellowlegs**

non-breeding

breeding

**Spotted Redshank**

non-breeding

**Common Greenshank**

non-breeding

breeding

**Red Knot**

juvenile

non-breeding

first-year adult

## Wood Sandpiper *Tringa glareola* 20cm

Small and graceful with rather short legs and short, straight bill. Wings reach tail tip. In flight, rump patch but no wing-stripe. Rather like Solitary Sandpiper but has dull yellow legs, paler upperparts and much more clearly defined white line above eye, with narrower orbital ring. Non-breeding adult plainer with buff-brown wash to throat and breast, but little streaking. Mantle sooty-brown with fewer white spots. Immature recalls non-breeding adult but has more prominent white supercilium, producing distinctly capped effect, dense but diffuse brown breast and upper belly streaking, warmer brown back and wing-coverts, spotted buff and cream. **Voice** A sharp, *pee pee peer*. **SS** From Lesser Yellowlegs by smaller size, shorter neck and wings, contrasting cap, longer supercilium and bill colour. **Status** Accidental Tobago, Buccoo, Dec 1996.

## Solitary Sandpiper *Tringa solitaria solitaria* 20cm

Sleek and graceful. Wings project just beyond tail. Flickering wingbeats. Breeding adult has blackish-brown head, throat, foreneck and breast, densely streaked white with conspicuous white orbital ring. Rest of underparts white. Mantle and rump blackish-brown, spotted white; flight feathers uniform black. Underwing black. Central tail feathers black, rest white with black bars. Bill black, legs greenish-grey. Non-breeding adult and immature similar but lack breast streaking and less spotted above. Two races, *solitaria* and *cinnamomea*, are virtually inseparable in winter plumage; *cinnamomea is* slightly larger and a warmer brown. **Voice** Sharp *peet-weet-weet*. **SS** From Wood Sandpiper by conspicuous orbital, darker head and breast markings, and dark central rump and tail. From non-breeding Spotted Sandpiper by longer bill, legs and wings, darker face and upperparts, darker underwing, leg colour and lack of supercilium and wing-stripe. **Status** Common non-breeding visitor to inland wetlands on Trinidad, less common on Tobago. Most records Aug–May.

## Spotted Sandpiper *Actitis macularius* 19cm

Short-necked and pot-bellied with wing-tips falling short of tail tip. Distinctive 'rear end' bobbing jizz. Fluttering stiff wingbeats followed by regular short glides, showing wing-stripe but no rump patch. Dark greyish-brown above with black barring and narrow white wing-bar in flight. Orbital ring, lower face and entire underparts white, but breeder is black-spotted below. Bill dull red with black tip, legs straw-yellow. Non-breeder and first-winter have narrow white bars, as well as black, on wing-coverts. **Voice** High-pitched whistle, *twee twee twee*. **SS** Breeder unmistakable. Other plumages from Solitary Sandpiper by shorter bill, wings and legs, paler head, upperparts and underwing, and presence of supercilium and wing-stripe. **Status** Common and widespread non-breeding visitor to wetlands across T&T. Present year-round but most in Aug–Apr.

## Willet *Catoptrophorus semipalmatus semipalmatus* 38cm

Tall and rather dumpy with mid-length, straight, thick bill. Rounded wings reach tail tips. In flight, both wing-stripe and rump patch. Greyish-brown above, streaked black on forehead to mantle, back and wing-coverts with black chevrons, rest of wings and tail barred blackish, primaries blackish and rump white. Bold white lores and eye-ring and underparts white streaked black on foreneck, becoming chevrons on breast, flanks and undertail-coverts. Non-breeder almost uniform grey above and whitish below. First-winter like pale adult, with broader pale fringes to upperparts. Two races probably occur, with *inornatus* paler and less heavily marked. **Voice** Ringing *kee-haa-haa* and a hoarse *wee-it wee-it*. **SS** From godwits by much shorter bill. **Status** Common non-breeding visitor, mainly to coastal mudflats of W Trinidad. Present year-round, but most in Aug–Nov. Less common on Tobago.

## Terek Sandpiper *Xenus cinereus* 23cm

Small with disproportionately long, slightly upturned bill, yellow with dark tip, horizontal gait and short yellow legs. Wings reach tail tip. Probes mud or scythes bill through water. Greyish-brown above with paler fringes, white below with fine dark streaks on face, neck and breast-sides. Broad white trailing edge obvious in flight. Breeders have dark scapulars and white fringes to upperparts. Immature recalls non-breeding adult but is generally darker grey-brown above with buff feather fringes. **Voice** Rapid whistling *wit-a-wit-wit*. **SS** Unmistakable. Occasionally pumps rear end while feeding, like Spotted Sandpiper. **Status** Accidental W Trinidad, Jun 1999.

**Wood Sandpiper**

non-breeding

**Solitary Sandpiper**

non-breeding

breeding

**Spotted Sandpiper**

juvenile

non-breeding

breeding

**Willet**

*inornatus*

*inornatus*

juvenile

non-breeding

*semipalmatus*

breeding
early spring

breeding
early spring

**Terek Sandpiper**

### Sanderling *Calidris alba rubida* 19cm

Small and robust with horizontal gait and short, straight bill. Wing-tips reach tail tip. Frenetic scampering, chasing the waves along beaches. Non-breeder has very pale crown and upperparts with browner wings fringed grey, and white trailing edge appears as part of longer wing-bar in flight; flight feathers and tail dusky. Bare parts all black. Breeder has irregular and very variable cinnamon to chestnut wash to head and back. Juvenile has some black on back. **Voice** A sharp *tik*. **SS** Unlikely to be confused. **Status** Fairly common visitor to coasts of E Trinidad and Tobago; Sep–Apr. Rarely, if ever, found inland.

### Semipalmated Sandpiper *Calidris pusilla* 16cm

Very small with short, broad-based and slightly 'blob-tipped' bill. Wing-tips reach tail tip. In flight, wing-stripe but no rump patch. Dainty pecking feeding action. Non-breeder warm grey above with fine streaks, broad white supercilium, slightly darker wings with white covert tips and blackish tail. Breeder washed brownish on head and back. Juvenile darker and browner. Bill and legs black. **Voice** A rough, hoarse *quiip* and lower-pitched twittering than Western. **SS** Very difficult to separate from Western Sandpiper. At all ages, assessing bill-length and shape and feeding action are good indicators. Breeding adult Semipalmated lacks rufous on head and scapulars. Non-breeder has greyer breast; immature has darker crown and duller buff-fringed scapulars. **Status** Abundant non-breeding visitor to wetlands across Trinidad; slightly less numerous Tobago. Some year-round, but most Jul–Apr.

### Western Sandpiper *Calidris mauri* 17cm

Very small with long, thin, slightly drooping bill. Wing-tips reach tail tip. In flight, wing-stripe but no rump patch. Distinctive probing feeding action. Non-breeder brownish-grey above with fine dark streaks from crown to wings, white forehead, supercilium and underparts, finely streaked on breast-sides. Bill and legs black. Breeder has some rufous on head and is more heavily spotted below. Immature darker and browner than breeder. **Voice** Single, sharp *cheep*, higher pitched than Semipalmated Sandpiper and distinctly longer. **SS** See Semipalmated. Breeding adult Western separable by rufous on crown, ear-coverts and upper scapulars. Non-breeders moult earlier, so in Jul–Oct appear whiter below. Most immatures are in first-winter plumage when they reach T&T. **Status** Abundant non-breeding visitor to mudflats on Trinidad, mostly Jul–Apr. Generally uncommon Tobago.

### White-rumped Sandpiper *Calidris fuscicollis* 19cm

Rather small, with attenuated jizz and mid-length, slightly decurved bill. Wings project well beyond tail. In flight, wing-stripe and conspicuous white uppertail-coverts. Non-breeder warm grey above with fine dark streaks on head and wings. Long, broad supercilium, chin and underparts white, streaked brown on breast, becoming vestigial at sides. Bill and legs black. Breeder darker above and streaked to upper belly and flanks. Immature has some chestnut wash on crown and back. **Voice** High thin *djeet* or *trrrt*. **SS** From Baird's by obvious rump patch. Appears heavier billed and larger headed. Non-breeding White-rumped greyer and plainer with bolder supercilium, breeder and immature much brighter. **Status** Uncommon passage migrant to inland and occasionally coastal wetlands of Trinidad; scarce on Tobago. Most Apr–May and Aug–Oct.

### Least Sandpiper *Calidris minutilla* 15cm

Tiny and drab with thin, finely pointed bill. Wings fall short of tail tip. In flight, wing-stripe but no rump patch. Distinctive crouched or hunched posture. Non-breeder has greyish-brown head, breast and upperparts, darker back and wings with paler fringes. Head and breast streaked dark brown. Bill black, legs greenish-yellow. Breeder more constrastingly marked. Immature rich brown above with bolder wing-bar and white outer scapular fringes. **Voice** A thin *prreep*. **SS** From all other small *Calidris* by tiny size, generally browner plumage and pale legs. **Status** Abundant non-breeding visitor, preferring inland wetlands on Trinidad; less numerous on Tobago. Some year-round, but most Jul–Apr.

### Baird's Sandpiper *Calidris bairdii* 19cm

Rather small and short-legged with attenuated jizz and mid-length straight bill. Wings project well beyond tail. In flight, wing-stripe but no rump patch. Non-breeder buffy-brown above with pale fringes, white supraloral separated by short eyebrow, pale cinnamon flush to breast (finely streaked darker). Short, finely pointed black bill, legs dusky-grey. Breeder darker and more contrasting above. Juvenile more heavily washed cinnamon. **Voice** A loud rolling *prreet*. **SS** From White-rumped by lack of white uppertail-coverts. Appears buffier with sleeker jizz, shorter legs, even longer wings, thinner, straighter all-black bill and lacks flanks streaking. **Status** Accidental Trinidad, Sep 1976, Nov 1989. Prefers grasslands over marshes or mudflats.

**Sanderling**

juvenile

breeding

non-breeding

**Semipalmated Sandpiper**

juvenile

non-breeding

**Western Sandpiper**

juvenile

non-breeding

**White-rumped Sandpiper**

white uppertail-coverts
show in flight

juvenile

non-breeding

**Least Sandpiper**

juvenile

non-breeding

**Baird's Sandpiper**

juvenile

non-breeding

## Pectoral Sandpiper *Calidris melanotos* 22cm

Medium-sized and pot-bellied with slightly decurved bill. Wings reach tail tip. In flight, wing-stripe and oval rump-side patches (central rump black). Non-breeder has warm grey head to breast, chestnut crown streaked dusky, and white throat and underparts. Back and wings to tail dusky or blackish, with buffy to grey fringes; central uppertail-coverts black, outers white. Breeder brighter with buff flush to breast and clearer supercilium. Immature bright, with some chestnut fringes above. Bill horn becoming dusky distally. **Voice** A hoarse, trilling *prrt*. **SS** From all other small sandpipers by clear-cut pectoral division. From Ruff by smaller size and slighter jizz. **Status** Common passage migrant to inland and occasionally coastal wetlands; often wet grassy fields on Trinidad. Less common Tobago. Mostly Aug–Oct, occasionally Apr–Jun.

## Curlew Sandpiper *Calidris ferruginea* 21cm

Rather tall with evenly decurved bill. Wings project slightly beyond tail. In flight, both wing-stripe and rump patch. Non-breeder has grey crown to wings, with pale grey fringes, flight feathers blackish with long white wing-bar and clear white uppertail-coverts in flight. Forehead and supercilium white, as are underparts, except grey flush to breast. Bill and legs black. Breeder has head and underparts, to belly, ruddy with clear white eye-ring and throat. Immature like non-breeder, but darker and browner above, streaked dusky. **Voice** Gentle *chirrup*. **SS** Breeding adult unmistakable. Other plumages from Stilt Sandpiper by bill shape, wing-stripe, whiter underwing and leg colour. **Status** Accidental Trinidad, May 2002.

## Stilt Sandpiper *Calidris himantopus* 22cm

Slender with upright posture, long, thick-based, distally drooping bill, and long legs. Wings project slightly beyond tail. In flight, rump patch but no wing-stripe. Appears like a yellowlegs but feeds like a dowitcher. Non-breeder has buffy-brown crown to wings, with a few dark feathers in upperparts. Complete white rump and uppertail-coverts, and white underparts with slight streaking on neck and scalloping to flanks and undertail-coverts. Bill and legs greenish-yellow. Breeder has bright orange face-sides and entire underparts scalloped grey and white, and dark upperparts. Immature more heavily marked than non-breeder. **Voice** A soft *cheoow*. **SS** Breeding adult unmistakable. See Curlew Sandpiper for separation of other plumages. From Short-billed Dowitcher by smaller size and slighter jizz, shorter bill and legs, shorter neck and lack of white rump. **Status** Common passage migrant to inland, and occasionally coastal, wetlands of Trinidad. Mostly Apr–May and Aug–Oct. Less common Tobago.

## Ruff *Philomachus pugnax* 22–30cm

Variable size with short, very slightly drooping bill. Small-headed, long-necked, large-bodied, pot-bellied and long-legged. In flight, wing-stripe and conspicuous white ovals on tail-sides. Non-breeding adult warm grey on crown and hindneck, with bold supercilium, darker back with pale grey fringes, and white wing-bar and rump-sides in flight. Underparts white, washed greyish on breast and sides. Juvenile has ochraceous buffy head to breast and flanks, rest of underparts white. Upperparts of young female dark with buffy fringes, blacker in male with pale grey to whitish fringes. **Voice** Usually silent. **SS** Breeding male unmistakable. All other plumages from Pectoral Sandpiper by larger size and lack of well-defined underparts markings. **Status** Rare passage migrant to inland wetlands in T&T, Apr–May and Aug–Jan.

## Wilson's Phalarope *Phalaropus tricolor* 23cm

Small-headed, long-necked and rather pot-bellied with needle-thin straight bill. In flight, obvious rump patch but no wing-stripe. Hyperactive, feeds on mud and often while swimming, picking insects off surface. Non-breeder grey above with fine white fringes, dusky flight feathers and white rump/uppertail-coverts. Underparts white washed greyish on breast-sides. Bill black, legs greenish-yellow. Juvenile/first-winter browner above. **Voice** A soft, nasal *aaugh*. **SS** From Stilt Sandpiper and Lesser Yellowlegs by feeding action, bill shape, face and underparts colour. **Status** Very rare visitor to inland wetlands in T&T, Aug–Oct.

## Buff-breasted Sandpiper *Tryngites subruficollis* 19cm

Long-necked, dome-crowned, with upright stance, short straight bill and wings projecting slightly beyond tail. Generally buffy, dark brown from nape to tail with broad pale fringes, and narrow white wing-bar. Bill dark, legs ochre-yellow. In flight, underwings white. Juvenile more densely spotted on head and neck. **Voice** Usually silent; occasionally a dry, soft *dreet* in flight. **SS** From young female Ruff by smaller size, shorter straighter bill, leg colour and lack of white in upperwing and tail. **Status** Rare passage migrant through T&T. Prefers grassy fields, golf courses, airfields and wet rice fields. Most records May, Sep–Oct.

**Pectoral Sandpiper**

juvenile

non-breeding

**Curlew Sandpiper**

juvenile

non-breeding

**Stilt Sandpiper**

juvenile

non-breeding

**Wilson's Phalarope**

juvenile/
first-winter

non-breeding

**Ruff**

♀
juvenile

♂
juvenile

♀
non-breeding

**Buff-breasted Sandpiper**

juvenile

non-breeding

# SKUAS AND JAEGERS – STERCORARIIDAE

Seven species form this worldwide family, which is characterised by aerial piracy, powerful flight, cryptic brown, grey and white body plumage and white flight-feather markings. Four species are confirmed for T&T waters, Long-tailed Jaeger is hypothetical and likely to occur. Behaviourally, Parasitics 'go for the ball', the other two 'go for the player!'.

## Great Skua *Stercorarius skua*     59cm

Broad-winged, bulky and short-tailed with powerful bill. Steady, slow and deep wingbeats, occasionally glides. Adult has dark brown head. Underparts brown densely streaked darker. Hindneck brown, densely streaked ginger. Mantle and rump brown, fringed buff. Upper- and underwing have brown coverts, fringed and tipped pale buff, with black flight feathers and large conspicuous white wing-flash. Tail brownish-black. Bill and legs black. Some are paler brown and amount of streaking above also very variable. Immature and subadult similar, but more reddish-brown with darker brown head, little if any underparts streaking, smaller white wing-flashes and two-toned bill. **Voice** Usually silent. **SS** From pale morph South Polar Skua by head and underparts colour. Rarely separable in field from dark morph though plumage warmer and browner. Immature from Pomarine Jaeger by heavier jizz, barrel-shaped body, shorter tail, darker upperparts, more pronounced white flashes on both wing surfaces, and lack of rump and flanks barring. **Status** One published report of a bird off Tobago but not admitted to the official checklist.

## South Polar Skua *Stercorarius maccormicki*     56cm

Powerful and bulky jizz, appearing hunched-backed, with very broad wings and short tail. Flies with steady, slow, deep wingbeats, occasionally gliding. Polymorphic. Pale morph adult has pale grey-buff head and underparts, palest on hindneck and bill surrounds, with an obvious dark eye. Back and wing-coverts dark grey-brown with variable buff fringes; flight feathers greyer with conspicuous white flash across base of primaries. Underwing as upperwing. Rump and tail dark grey-brown. Bill and legs black. Dark morph adult wholly brownish-black with slightly paler bill surrounds and yellowish nape stain. Both upper and underwing flight feathers have prominent white flash across base of primaries. There are also intermediate morphs. Immature less variable with cold dark grey or grey-brown body plumage slightly smaller white wing-flashes, blue-grey bill, tipped black and bluish legs. **Voice** Usually silent. **SS** Pale morph adult unmistakable. Other morphs rarely separable at sea from Great Skua. **Status** Accidental to Trinidad coastal waters. One adult pale morph off Icacos Pt, July 1980. Two further unidentified skuas reported from Tobago, Mar 1986 and Galera Pt, Trinidad, Nov 2002.

**Great Skua**

juvenile
pale morph

adult average
intermediate
morph

juvenile
dark morph

adult
pale morph

adult average
intermediate
morph

adult
dark morph

juvenile
intermediate
morph

**South Polar Skua**

adult
pale morph

adult
intermediate
morph

adult
dark morph

adult pale
intermediate
morph

adult dark
intermediate
morph

adult
dark morph

## Long-tailed Jaeger *Stercorarius longicaudus longicaudus* 38cm

Slender, almost dainty and gentle in appearance, with short bill, long narrow wings and attenuated shape. Flight buoyant, with rather weak wingbeats followed by long glides and shearing. Breeding adult very neat with small black cap and foreface. Throat, lower and rear face creamy white, continuing as complete nuchal collar. Neck-sides, breast and belly white, becoming brownish-grey on rear flanks and undertail-coverts. Mantle, rump, wing-coverts and inner flight feathers cold, pale mousy grey-brown with conspicuous black trailing edge. Outer flight feathers black with just two visible white outer primary shafts. Underwings uniform grey-brown. Tail black with two extremely long central rectrices, but often difficult to see at distance. Bill black, legs grey. Non-breeding adult has patchy black cap, diffuse grey-brown breast-band, white shaft on underside of outermost primary, heavily barred rear flanks, undertail-coverts, rump and uppertail-coverts, and lacks long central tail feathers. Juvenile highly variable, with pale and dark morphs and many intergrades, but always appears slim and attenuated with slightly round-tipped longer central rectrices. All are cold-coloured, from pale grey to black. Underparts vary from grey-white to grey-black with heavily barred undertail-coverts. Most have pale mantle and wing-coverts fringes, darker flight feathers with 2–3 white primary shafts, heavily barred underwing with narrow white wing-flash, heavily barred rump and blackish tail. Bill basally grey-white, distally black. Legs blue-grey. **Voice** Usually silent. **SS** From Parasitic and Pomarine Jaegers by smaller, slighter structure, and more buoyant flight. Adult by neat appearance and distinctive dark upperwing trailing edge. Immature is colder with a more rounded head, thinner bill, fewer white upperwing primary shafts and smaller white underwing flash. **Status** Yet to be recorded in T&T waters, but recorded off mainland; mostly Nov–Feb.

## Parasitic Jaeger *Stercorarius parasiticus* 45cm

Gull-like but sleek, with long, thin angled wings. Light, agile flight with quick wingbeats, interspersed by glides, twists and turns. Dimorphic with intermediates. Breeding adult pale morph has black crown and upper face giving capped appearance, creamy buff lower face and throat continuing as nuchal collar and merging into off-white lower breast and belly. Diffuse dark brown neck-sides patches occasionally form upper-breast-band. Lower belly and undertail-coverts smoky brown. Back and wing-coverts dark brown, flight feathers black with 3–6 distinct white primary shafts. Underwing dark brown with white flash at base of primaries. Rump and tail dark brown with pointed projecting central rectrices. Bill thin and black with pale surround; legs black. Breeding dark morph adult dark brown with similar white flight-feather markings. Non-breeding adult similar but has less well-defined cap, pale mantle fringes and lacks central tail projections. Immature varies from dark ginger to blackish-brown, faintly barred black. Flight-feather markings as adult, tail blunt, with two, slightly protruding pointed central tail feathers. Bill basally blue-grey, distally black; legs blue-grey. **Voice** Usually silent. **SS** At all ages, from Pomarine by flight action, wing, body and bill shape and size. Breeding adult by tail-length and shape. Some immatures not safely separated, but note faint underwing-coverts barring, dark uppertail-coverts and pointed central tail feathers. **Status** Scarce passage migrant, mostly Dec–Apr, Sep–Oct.

## Pomarine Jaeger *Stercorarius pomarinus* 56cm

Large and bulky, heavy-billed, pot-bellied and broad-winged. Deep ponderous wingbeats and occasional glides; rarely shears. Dimorphic with intermediates. Breeding adult pale morph has black head and nape, with creamy buff throat and lower ear-coverts continuing as nuchal collar, separated from off-white lower breast and belly by dark brown diffuse, and sometimes incomplete, upper-breast-band. Upper flanks heavily barred dark brown, undertail-coverts solid brown. Back and wing-coverts dark brown, flight feathers blackish with 5–8 distinct white primary shafts. Underwing dark brown with white flash at base of primaries. Rump and tail dark brown with very long central rectrices twisted like two spoons. Bill distinctly two-toned, basally yellowish-brown, distally black. Legs blackish. Dark morph adult much rarer, brownish-black with similar white flight-feather markings. Non-breeding adult dirtier with pale upperparts fringes, heavily barred rump, upper- and undertail-coverts, and lacks central tail 'spoons'. Immature has cold, dark grey-brown plumage with variable pale fringes, strongly barred both underwing- and uppertail-coverts, blunt, rather rounded tail and flight-feather markings as adult. **Voice** Usually silent. **SS** At all ages from Parasitic Jaeger by different flight, bulk, bill size and broader-based wings. Breeding adult by elongated central tail feathers. Most immatures by colder body plumage lacking rufous or ginger tones, strong underwing- and uppertail-coverts barring and rounded central tail feathers. **Status** Rare passage migrant to coastal waters in T&T: just six birds in last 12 years, all Jan–Apr.

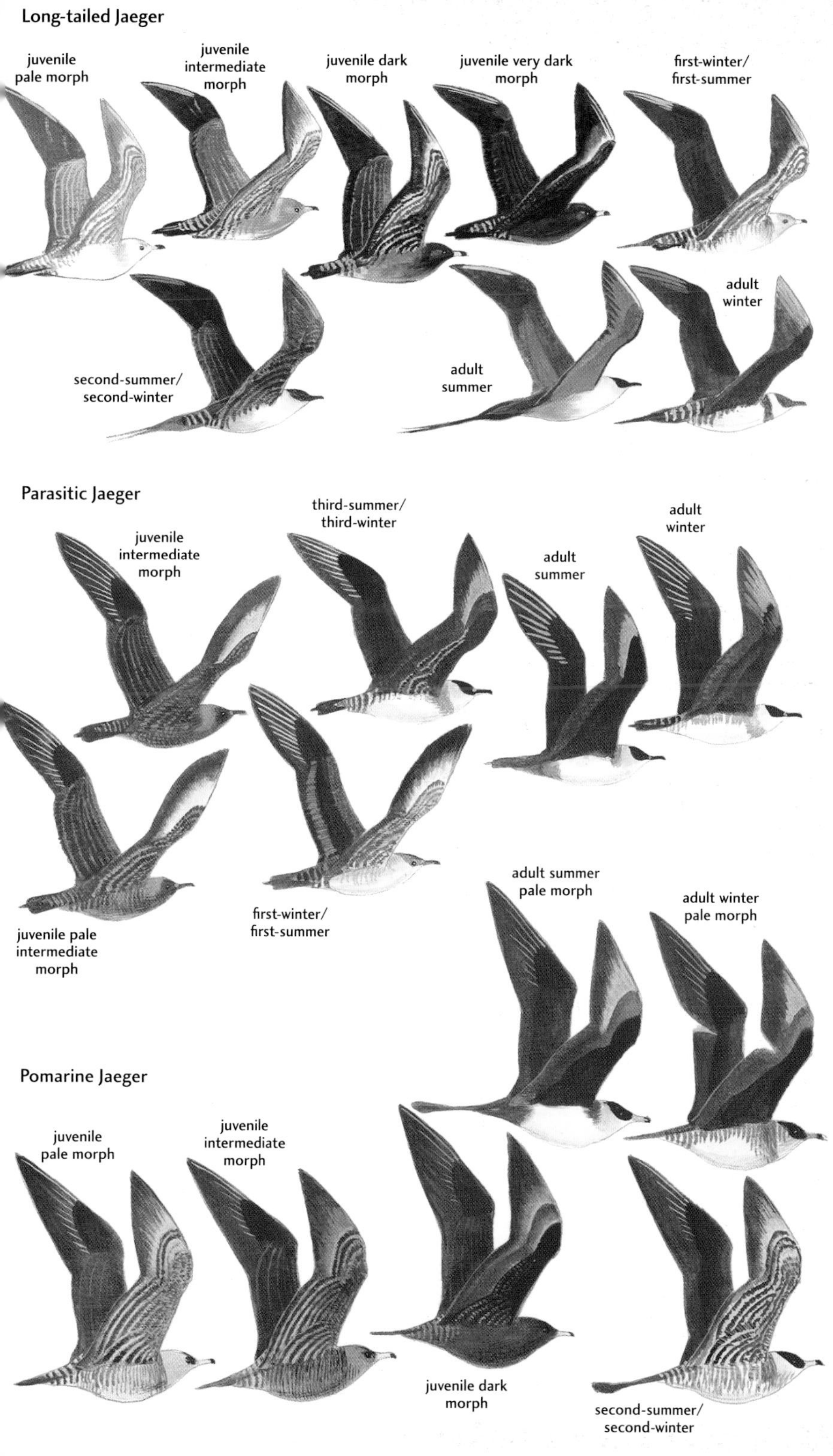

# Long-tailed Jaeger

juvenile
pale morph

juvenile
intermediate
morph

juvenile dark
morph

juvenile very dark
morph

first-winter/
first-summer

second-summer/
second-winter

adult
summer

adult
winter

# Parasitic Jaeger

juvenile
intermediate
morph

third-summer/
third-winter

adult
summer

adult
winter

juvenile pale
intermediate
morph

first-winter/
first-summer

adult summer
pale morph

adult winter
pale morph

# Pomarine Jaeger

juvenile
pale morph

juvenile
intermediate
morph

juvenile dark
morph

second-summer/
second-winter

# GULLS AND TERNS – LARIDAE

Gulls take up to four years to attain adulthood and exhibit distinct plumage maturation. Unlike in temperate climates, gulls in T&T are rarely found inland. Terns attain adult plumage within a single calendar year. Of 23 species occurring in T&T just two are resident.

## Ring-billed Gull *Larus delawarensis*                                    45cm; WS 122cm

Medium-sized, bulky three-year gull with short, straight, thick bill. Non-breeding adult has pearl grey back and wings, primary tips black with white spot, trailing edges white; rest white with slight brown streaking on nape. Bill yellow with black subterminal band, legs yellow. Breeder has all-white head and warm yellow legs. From juvenile to third-winter gradual progression from very variable, streaked brown head and underparts, dark wings with pale fringes, and dark subterminal tail-band, to much as non-breeder, but with marks on primary coverts, outer secondaries and greater coverts, and vestigial tail-band; bill pinkish with black tip and legs pinkish in all subadult plumages. **Voice** A hoarse wheezy cry. **SS** Unlikely to be confused. **Status** Rare visitor to coastal W Trinidad. Most records Nov–Mar. No recent sightings Tobago.

## Kelp Gull *Larus dominicanus dominicanus*                                58cm; WS 135cm

Bulky, powerful four-year gull with broad, rounded wings and heavy, droop-tipped bill. Non-breeder has black back and upperwings, with leading and trailing edges white, and grey underside to flight feathers. Some pale brownish streaking on head, but not noticeable at distance. Bill yellow, sometimes with vestigial red gonys spot; legs dull greenish-yellow. Breeder has all-white head and more obvious gonys spot. Juvenile well streaked and spotted mid-brown, wings, back and tail dark brown; bill dark brown, legs greyish-pink. Intermediates gradually become paler, with central rectrices becoming white first. **Voice** A hoarse, loud *kee-aa*. **SS** Appears larger, bulkier and longer necked with more robust and powerful bill than American Herring or Lesser Black-backed Gulls. Adults and subadults further separated by mantle and underwing colour; immature from American Herring Gull by head and underparts mottling. **Status** Accidental to coastal W Trinidad: an adult at various sites, Jul 2000–Jan 2001.

## American Herring Gull *Larus smithsonianus*                              64cm; WS 147cm

Large, four-year, long-winged gull with powerful bill and sloping forehead peaking behind eye. Considerable individual variation in immatures and subadult. Non-breeder has grey upperwings and back, with black tips to outermost five primaries and a few white subterminal spots; leading edge to inner wing and entire trailing edge white. Underwing often slightly greyish with weak blackish tips to outer primaries. Brownish streaks on head and neck. Bill yellow with red gonys spot; legs pink. Breeder has all-white head. Juvenile brown with blackish bill and pinkish-grey legs. Intermediates gradually gain white head, tail base, rump and underparts. **Voice** A loud piercing *keeah*. **SS** From adult Lesser Black-backed Gull by mantle, wing and leg colour. Immature by boldly patterned tertials, inner primary 'window' and contrasting pale rump. **Status** Accidental Trinidad; just three records, none since 1982.

## Lesser Black-backed Gull *Larus fuscus graellsii*                        53cm; WS 137cm

Large, four-year gull. Sleek with rounded head and long, narrow wings. Non-breeder has wings and back black, with leading edge to inner wing very narrowly white and trailing edge entirely white. Underside to flight feathers grey with white trailing edge. Variable brown streaking on head and hindneck. Bill yellow with red gonys spot; legs pink. Breeder has all-white head, soft parts richer. Juvenile streaked brown on head and body, but paler on belly and tail-coverts; bill blackish, legs pinkish-grey. Intermediates to third-winter attain white rump and tail-coverts, back becoming grey first. **Voice** A guttural *kow*. **SS** From adult American Herring Gull by mantle, wing and leg colour. Immature by duller tertial pattern, inner primaries colour, broader dark upperwing trailing edge and slightly darker rump. **Status** Locally uncommon non-breeding visitor to mudflats of W Trinidad; 1–2 present year-round. Rare on Tobago.

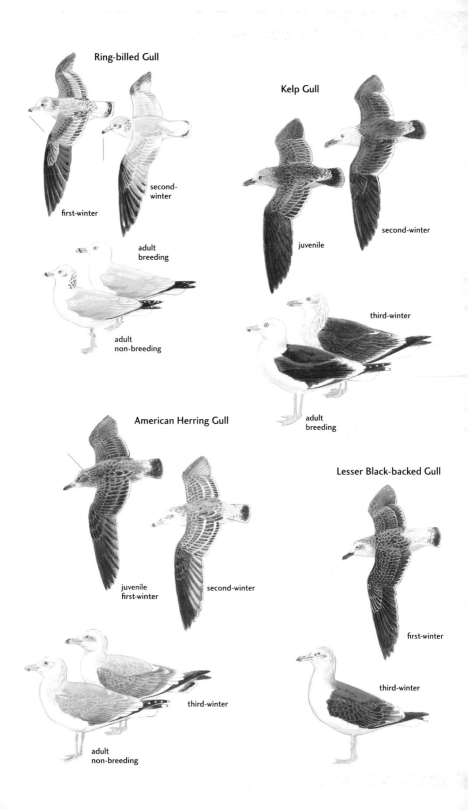

**Ring-billed Gull**

first-winter

second-winter

adult breeding

adult non-breeding

**Kelp Gull**

juvenile

second-winter

third-winter

adult breeding

**American Herring Gull**

juvenile first-winter

second-winter

third-winter

adult non-breeding

**Lesser Black-backed Gull**

first-winter

third-winter

## Laughing Gull *Larus atricilla atricilla*      42cm; WS 102cm

Medium-sized, three-year gull with long droopy bill, sloping forehead and very long wings. Breeding adult has complete black hood with bold white eye-crescents. Underparts and nape white. Mantle and upperwing dark grey with black wing-tips, three small white primary tip spots and white trailing edge to inner flight feathers. Underwing-coverts dirty white, inner flight feathers grey, rest black. Rump and tail white. Bill dark red with black subterminal band, legs dark red. Non-breeder similar but head white with grey smudge on hindcrown and ear-coverts. Bill and legs black. Subadult recalls non-breeder but has grey wash to upper breast and flanks, grey nape, no visible wing-tip spots and partial black terminal tail-band. First-year from subadult by extensive grey mottling on head and underparts, nape and mantle mid-grey, wing-coverts brownish-grey, primary-coverts and flight feathers black. Underwing-coverts grey, smudged darker, flight feathers black. Tail feathers broadly tipped black. Immature from first-winter by brownish-grey head and breast, rest of underparts white and brownish-grey mantle. **Voice** A nasal penetrating laugh. **SS** From Franklin's Gull at all ages by larger size, heavier drooping bill, wing shape and length, and underwing colour. Adult and subadult by different upperwing markings; immature by head markings and complete tail-band. **Status** Abundant resident on coasts of W and S Trinidad. In Nov–Mar roosts numbering several thousands form on mudflats. Uncommon elsewhere. Common resident coastal Tobago with large breeding colony on Little Tobago.

## Franklin's Gull *Larus pipixcan*      37 cm; WS 91 cm

Dainty two-year gull with domed crown, short straight bill and rounded short wings extending just beyond tail. Breeding adult has complete black hood with bold white eye-crescents. Underparts white with pink flush. Nape white, mantle, wing-coverts and basal half of flight feathers dark grey with broad white trailing edge. White band separates black outer feathers from 3–4 large white tips. Underwing very pale grey with trailing edge and wing-tips as above. Rump and tail white. Bill bright red, legs darker. Non-breeder similar but has white forehead and lores, extensive black rear hood, black bill and legs. First-winter recalls non-breeding adult but has brownish-grey upperwing-coverts, narrower trailing edge and black outer flight feathers. They lack both white subterminal band and white wing-tips, and have an incomplete black terminal tail-band. **Voice** A querulous far-carrying *koyu*. **SS** From Laughing Gull at all ages by smaller more delicate profile, different head, bill and wing shapes, and pale underwing. Adult and subadult by broad white trailing edge, white subterminal wing-bar and obvious wing-tips; immature by darker hood and partial tail-band. **Status** Very rare visitor to coasts of W Trinidad: just seven documented records in last 12 years, all Dec–Apr. No sightings on Tobago.

adult
breeding

adult
non-breeding

first-summer

juvenile

first-winter

first-winter

Laughing Gull

adult
non-breeding

adult
breeding

Franklin's Gull

juvenile

second-winter

first-winter

adult
non-breeding

first-summer

first-winter

juvenile

## Black-headed Gull *Larus ridibundus*                41 cm; WS 102 cm

Medium-sized two-year gull with long, broad wings. Breeding adult has chocolate-brown hood not extending onto nape, with white eye crescents. Nape and underparts white. Mantle, wing-coverts and inner flight feathers pale grey. Outer flight feathers white, tipped black, forming pronounced white leading edge. Underwing flight feathers dusky, outermost white. Rump and tail white. Bill and legs dark red. Non-breeder similar but has white head with black ear smudge. Bill and legs brighter red. First-winter recalls non-breeding adult but has brown carpal bar, broad black trailing edge to flight feathers, broad black terminal tail-band, orange bill broadly tipped black, and orange legs. **Voice** Harsh *grreek*. **SS** Unlikely to be confused. **Status** Rare visitor to coasts of T&T. Seven birds in last 12 years, Nov–Jul.

## Black-legged Kittiwake *Rissa tridactyla tridactyla*                43 cm; WS 91 cm

Two-year gull, long-winged and short-legged with rounded crown. Breeding adult has white head with black eyes. Nape and underparts white. Mantle and wing-coverts dark grey. Rump and tail white. Flight feathers grey-white with black wing-tips. Bill yellow, legs black. Non-breeder similar but has dusky crescent-shaped ear-spot. First-winter has blacker ear-spot, black nape collar, M-shaped black lines formed by tips of wing-coverts and leading edge of outer flight feathers, black terminal tail-band and black bill. **Voice** Usually silent. **SS** Adult unmistakable. Immature from Sabine's Gull by larger size, longer wings, paler mantle and wing-coverts and M-shaped black upperwing markings. **Status** Accidental to coast of N Trinidad: just one record, Feb 1998.

## Sabine's Gull *Xema sabini*                34 cm; WS 84 cm

Two-year gull. Appears tiny at rest with small head and short bill. In flight, diagnostic wing pattern, pointed wings, shallow-forked tail and graceful tern-like action. Breeding adult has grey-black head. Nape and underparts white. Mantle grey, rump and tail white. Upperwing forms three clear triangles; wing-coverts dark grey, secondaries and inner primaries white, outer primaries black. Underwing white with dusky bar on greater coverts and black outer primary tips. Bill black with yellow tip, legs black. Non-breeder similar but has black rear crown and ear-coverts, and thin white eye crescents. Immature differs by white forecrown, lores and throat; rest of crown and face grey-brown. Nape, mantle, neck-sides patch and wing-coverts brownish-grey, fringed white. Tail white with black terminal band. Bill black, legs pinkish-grey. **Voice** A harsh, grating trill. **SS** Adult unmistakable. Immature from Black-legged Kittiwake by smaller size and darker head, nape, mantle and wing-coverts. **Status** Accidental to Trinidad, with two records, Mar 1982 and Jan 2002.

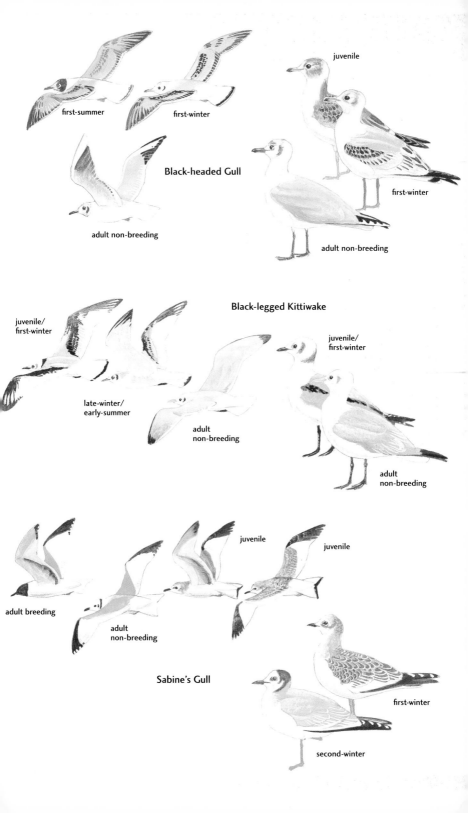

**Black-headed Gull**

first-summer

first-winter

juvenile

first-winter

adult non-breeding

adult non-breeding

**Black-legged Kittiwake**

juvenile/
first-winter

juvenile/
first-winter

late-winter/
early-summer

adult
non-breeding

adult
non-breeding

juvenile

juvenile

adult breeding

adult
non-breeding

**Sabine's Gull**

first-winter

second-winter

## Gull-billed Tern *Gelochelidon nilotica aranea* 35cm; WS 100cm

Medium-sized tern, broad- and round-winged, heavy-bodied and short-necked, with deep-based, stout bill and short, slightly forked tail. Does not plunge-dive, but leisurely hawks insects. Breeding adult has black crown, nape and face to eye level. Lower face and underparts white. Mantle and wings pale grey with dusky shafts and trailing edge to outer primaries. Underwing white with black outer primary trailing edge. Rump and central tail pale grey, outer tail white. Bill and legs black. Non-breeder and first-winter have white head with black eye-patch. **Voice** A hoarse *kee wak*. **SS** When breeding, from Sandwich Tern by heavier body, more rounded wings, bill structure, bill, rump and tail colour. Also note white head and black eye-patch. **Status** Uncommon non-breeding visitor to inland and coastal waters of W Trinidad; rare on passage off NE coast. Most sightings early Aug–mid Apr. No records from Tobago.

## Caspian Tern *Hydroprogne caspia* 55cm; WS 135cm

Very large; bulky with deep-based, dagger-like bill and slightly forked tail. Breeding adult has black crown and upper face to eye level. Lower face, nape and underparts white. Mantle and upperwing pale grey with dusky trailing edge to outer primaries. Underwing white with dark outer primaries. Rump and tail white. Bill coral-red, tipped black, legs black. Non-breeder has white streaking in black crown and upper- and underwing primaries extensively blackish. Immature recalls non-breeding adult but has grey-brown mantle mottling, greyer upperwing flight feathers and darker tip to outer tail. **Voice** A deep, harsh scream. **SS** From Royal Tern by larger size, heavier and bull-necked build, bill shape and colour, and dark underside to flight feathers. In all non-breeding plumages by dark forehead. **Status** Accidental to coast of W Trinidad. Just two records, most recent at Orange Valley, Nov 2005.

## Royal Tern *Thalasseus maximus maximus* 51cm; WS 130cm

Large but sleek with thick, straight bill and slender wings that extend well beyond deeply forked tail. Breeding adult has black, shaggy crown, upper nape and face extending to eye level. Lower face, nape and underparts white. Mantle, wing-coverts and inner flight feathers pale grey, outer flight feathers fringed and tipped black. Underwing white with just tips of outer flight feathers black. Rump and tail white. Bill orange-red, legs black. Non-breeding adult similar but has white forecrown, black rear crown, nape and ear-coverts. Outer flight feathers darker through wear. First-winter from non-breeding adult by brown flecking on mantle. Upperwing has black inner wing leading and trailing edges and outer flight feathers. Outer tail feathers dusky. Bill and legs orange-yellow. **Voice** A shrill *hee-arr*. **SS** From Caspian Tern by smaller size, narrower wings, bill size and colour, underwing pattern and more deeply forked tail. From Cayenne Tern by larger size and bill colour. **Status** Common visitor to coasts of both islands; has bred. Present year-round, but numbers greatest Nov–Apr.

## Large-billed Tern *Phaetusa simplex simplex* 38cm

Large with bold, diagnostic wing pattern, broad-based, long, slightly drooping bill and slightly forked tail. Breeding adult has black crown and face to eye level; nape, lower cheeks and throat white. Underparts dirty white. Mantle and wing-coverts mid-grey, inner flight feathers white and outer ones black, forming three distinct triangles of colour. Underwing mainly white. Rump and tail darker grey. Bill bright yellow, legs yellowish-green. Non-breeder has white-mottled forehead, crown and nape. Immature has pale grey crown, streaked black, black eyeline, pale grey flanks and brownish tips to upperparts. **Voice** Raucous *kai aat*. **SS** Unmistakable. **Status** Common visitor to wetlands of C and W Trinidad; occasionally breeds. Present Feb–Nov. No documented records for Tobago.

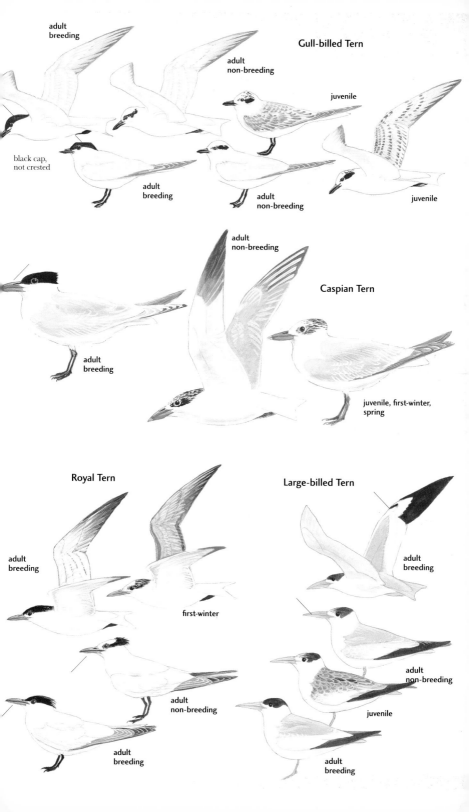

adult
breeding

Gull-billed Tern

adult
non-breeding

juvenile

black cap,
not crested

adult
breeding

adult
non-breeding

juvenile

adult
non-breeding

Caspian Tern

adult
breeding

juvenile, first-winter,
spring

Royal Tern

Large-billed Tern

adult
breeding

adult
breeding

first-winter

adult
breeding

adult
non-breeding

adult
non-breeding

juvenile

adult
non-breeding

adult
breeding

adult
breeding

## Sandwich Tern *Thalasseus sandvicensis acuflavidus* 38cm

Mid-sized, very white tern. Slender wings with tips reaching short, deeply forked tail tip. Appears front-heavy in flight with deliberate, deep wingbeats. Breeding adult has black shaggy crown, upper nape and face to eye level. Lower face, lower nape and underparts white. Mantle, wing-coverts and inner flight feathers very pale grey, rest dusky-grey. Underwing white with dusky trailing edge to outer primaries. Rump and tail white. Bill black with yellow tip, legs black. Non-breeding adult has white lores and forehead. First-winter recalls non-breeding adult, with dark centres to wing-coverts and tertials and dusky bars on leading and trailing edges of inner flight feathers. **Voice** Loud, scratchy *kirrik*. **SS** From Gull-billed Tern by wing and tail shape, rump colour, bill shape and colour. **Status** Uncommon visitor to coasts of T&T. Recorded year-round but most in Apr–Nov.

## Cayenne Tern *Thalasseus [sandvicensis] eurygnatha* 38cm

Appears identical to Sandwich Tern in overall size, shape and plumage colour. Indeed, they are known to hybridise and are currently considered conspecific by many authorities. Bill colour is extremely variable; most are greenish-yellow, sometimes tinged orange or red; others show variable amounts of black. **Voice** Shrill *keerak*. **SS** From Royal Tern by smaller size, longer crest when breeding (but note that black on front of head starts to moult out as soon as the first egg is laid), longer tail and thinner, mainly greenish-yellow bill. **Status** Uncommon breeding visitor to coasts of T&T. Recorded year-round but most in Apr–Nov.

## Roseate Tern *Sterna dougallii dougallii* 38cm

Mid-sized tern that appears white-winged in flight. Tail extends well beyond wing-tips. Shallow, stiff and rapid wingbeats. Breeding adult has black crown, nape and upper face to eye level; lower face white. Underparts white. Mantle and upperwing grey-white with outermost three flight feathers black. Underwing white with translucent trailing edge. Rump and deeply forked tail white with extremely long narrow outer tail-streamers. Bill dark red, paler at base, legs bright red. Non-breeding adult has white forehead. First-winter recalls non-breeding adult, with black and brown mantle barring, dark upperwing leading edge contrasting with white trailing edge, and black legs. **Voice** A soft *chivik* and a hoarse *kaah*. **SS** From Common Tern at all ages by paler upperparts, black restricted to outermost upperwing flight feathers, white underwing, longer all-white tail, darker bill and different flight action. **Status** Common breeding visitor to coastal Tobago, but uncommon on passage off coasts of N and E Trinidad. Most records Apr–Oct.

## Common Tern *Sterna hirundo hirundo* 35cm

Mid-sized tern. Wing-tips extend slightly beyond tail. Buoyant flight with powerful wingbeats. Breeding adult has black crown, nape and face to eye level; lower face white. Underparts grey-white. Mantle and upperwings pale grey with dark grey outer flight feathers; where these meet, on trailing edge, blackish wedge points inwards. Underwing white with narrow black leading edge and broader black trailing edge to outer flight feathers. Inner primaries and outer secondaries appear translucent. Rump and deeply forked tail white. Bill reddish-orange, tipped black, legs red. Non-breeder has white forehead, dark leading edge carpal bar on upperwing and black bill. First-winter recalls non-breeding adult with dark grey secondaries and black outer primaries. **Voice** Shrill *kee-aaa*. **SS** From Roseate Tern at all ages by darker upperparts, more extensive black on upper- and underwing flight feathers, shorter tail, bill colour and more purposeful flight action. **Status** Common non-breeding visitor to coasts of T&T. Recorded in all months but most numerous on coasts of NE and S Trinidad Aug–Oct.

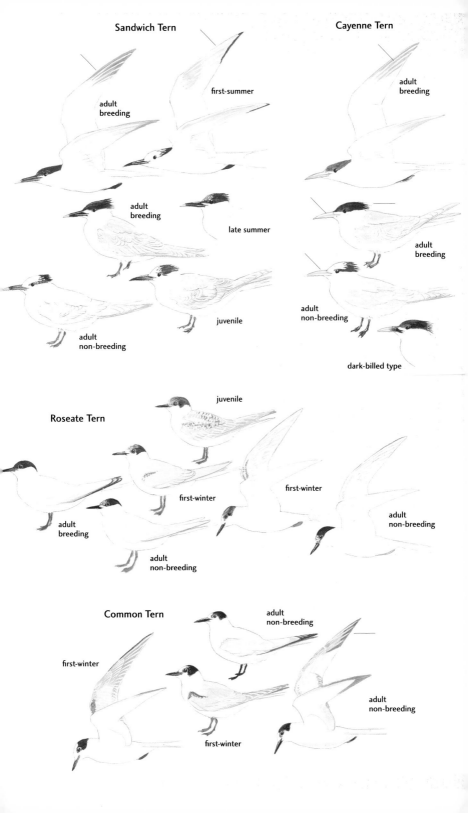

**Sandwich Tern**

adult
breeding

first-summer

adult
breeding

late summer

adult
non-breeding

juvenile

**Cayenne Tern**

adult
breeding

adult
breeding

adult
non-breeding

dark-billed type

**Roseate Tern**

juvenile

first-winter

first-winter

adult
breeding

adult
non-breeding

adult
non-breeding

**Common Tern**

adult
non-breeding

first-winter

first-winter

adult
non-breeding

### White Tern *Gygis alba alba* 30cm

Rather small and delicate with blunt-tipped wings and short, forked tail. Flies gracefully and buoyantly with deep floppy wingbeats. Adult has milky white plumage with large black eyes. Bill black, legs blue-grey. Immature has white forehead; rear crown dusky. Face white with black spot on ear-coverts. Underparts white. Back and wing-coverts pale grey with brown-mottled mantle. Flight feathers white with black shafts on outer feathers; rump and tail white. **Voice** Usually silent. **SS** Unmistakable. **Status** Accidental SW Tobago: just one record, Aug 1987.

### Bridled Tern *Onychoprion anaethetus recognita* 35cm

Long-winged with very long, forked tail. Graceful and erratic flight with buoyant wingbeats. Adult has white forehead and supercilium reaching slightly behind eye; rest of crown black. Face white with black lores and upper ear-coverts. Underparts white. Neck-sides grey, rest of upperparts darker grey-brown; central tail feathers grey, rest white. Wing-coverts grey-brown, flight feathers darker. Underwing-coverts white, flight feathers grey. Bill and legs black. Non-breeding adult has more white on crown. Immature has white forecrown; rear crown patchily grey. Face white with black ear-coverts smudge. Underparts white with buff wash to upper breast and neck-sides. Back and wing-coverts grey, tipped buff and white, flight feathers darker. Tail and underwing as adult. **Voice** Soft, nasal whine, *naark*. **SS** Adult from Sooty Tern by flight action, contrasting grey upperparts and black crown separated by pale collar. Tail feathers extensively white with just central ones grey. Longer white supercilium very difficult to discern in field. **Status** Uncommon breeding visitor to offshore islands of Tobago, late Apr–late Jul. Rare off NE coast of Trinidad.

### Sooty Tern *Onychoprion fuscatus fuscatus* 38cm

Long-winged with long, deep-forked tail. Flight purposeful and direct with powerful wingbeats. Breeding adult has white forehead which becomes short supercilium reaching eye; rest of crown black. Face white with black lores and upper ear-coverts. Underparts white. Upperparts black with just outer tail feathers white. Underwing-coverts white, flight feathers dark grey. Bill and legs black. Non-breeder has mottled crown and nape. Immature has head, breast, flanks and belly sooty-brown, undertail-coverts dirty white. Upperparts black with white mantle and wing-coverts spotting. Underwing-coverts white, flight feathers dusky. **Voice** Nasal *kee weekk weekk*. **SS** Adult from Bridled Tern by black upperparts, more extensive black in tail, and flight action. Difference in white on foreface only apparent at close range. Immature from similar Brown Noddy by white underwing-coverts. **Status** Abundant breeding visitor to offshore islands of Tobago, late Apr–late Jul. Occasionally seen off NE coast of Trinidad.

### Black Tern *Chlidonias niger surinamensis* 24cm

Small with broad wings and short, slightly forked tail. Very buoyant and erratic flight. Breeding adult has black head, throat, breast and belly; undertail-coverts white. Upperparts grey. Underwing-coverts dirty white, flight feathers grey. Bill black, legs dark red. Non-breeder has white forecrown with dusky wraparound rear-crown cap extending to rear ear-coverts. Rest of face and underparts white with grey wash and dusky breast-sides patch. Nape collar white, rest of upperparts and underwing grey. Immature recalls non-breeding adult, with smudgy grey flanks, brownish-grey mantle and darker carpal bar and flight feathers. **Voice** A harsh *keep*. **SS** From Sooty and Bridled Terns by much smaller size, short broad wings and different flight action. **Status** Rare passage migrant to NE coast of Trinidad, Aug–Oct. Very rare elsewhere.

White Tern

adult

juvenile

Bridled Tern

juvenile

adults

Sooty Tern

juvenile

adult

first-summer

adult

juvenile

adult

Black Tern

non-breeding
adult

first-summer/
winter

breeding adults

non-breeding
adult

### Least Tern *Sternula antillarum antillarum* 24cm

Small with slender bill and slightly forked tail; rapid wingbeats. Breeding adult has white forehead, black crown and nape, broad black stripe from bill to nape and short white supercilium. Lower face and underparts white. Upperparts pale grey with just outermost two flight feathers black. Underwing pale grey with darker outermost feathers. Bill yellow with black tip, legs yellow. Non-breeder has more extensive white forecrown and black bill. Immature recalls non-breeding adult, with mainly white crown, brown-mottled mantle, black carpal bar and black bill. **Voice** High-pitched *kdeet* and shrill *kit kit kit*. **SS** From breeding adult Yellow-billed Tern by black bill tip; from non-breeding adult by black bill and immature by bill colour and more pronounced carpal bar. **Status** Uncommon passage migrant off N and E Trinidad; scarcer off W coast. Most records Aug–Oct. Very rare on Tobago.

### Yellow-billed Tern *Sternula superciliaris* 23cm

Small with slender bill; rapid wingbeats. Breeding adult has white forehead, black crown and nape, broad black stripe from bill to nape and short white supercilium. Underparts white, upperparts pale grey with outermost three flight feathers black. Underwing pale grey with darker trailing edge and outer wedge. Bill yellow, legs pale yellow. Non-breeding adult has white crown, lores and nape flecked black, and dark tip to bill. Immature recalls non-breeding adult, with grey-mottled face and dark grey carpal bar. **Voice** Fast scratchy *ki ki ki ki ki*. **SS** From breeding adult Least Tern by all-yellow bill; from non-breeding adult by yellow not black bill, and immature by bill colour and less distinct carpal bar. **Status** Common non-breeding visitor to inland wetlands and mudflats of W Trinidad, Mar–Nov at least. Several undocumented reports from Tobago.

### Brown Noddy *Anous stolidus stolidus* 38cm

Long-winged with long, wedge-shaped tail. Flight erratic, with powerful wingbeats and often shears low over water. Adult sooty-brown with grey-white forecrown. Wing-coverts sooty-brown, flight feathers and tail darker. Underwing-coverts paler grey-brown with dark fringes. Bill and legs black. Immature recalls adult but lacks pale cap. **Voice** Silent at sea; on breeding grounds a guttural croak. **SS** Immature from same-age Sooty Tern by dark underwing-coverts; otherwise unmistakable. **Status** Common visitor to coastal Tobago, breeding on offshore islands, late Apr–early Aug. Usually scarce on NE Trinidad coast, but very common on passage Aug–Oct.

## SKIMMERS – RYNCHOPIDAE

Just three species occur worldwide, all of them similarly plumaged. Long wings project well beyond tail tips, unique bill shape and very short legs. Feed by skimming low over water, slicing surface with lower mandible. Just one species occurs in T&T.

### Black Skimmer *Rynchops niger cinerascens* 46cm; WS 112cm

Breeding adult has black crown and face to eye level. Forecrown, lower face and underparts white. Upperparts black with narrow white trailing edge to inner flight feathers and outer webs of outer tail feathers. Underwing dark grey. Bill basally red and distally black; upper mandible much shorter than lower. Legs coral-red. Non-breeder has whitish collar. Juvenile browner above with scaled mantle and scapulars, and orange-red bill. **Voice** A nasal *uurk*. **SS** Unmistakable. **Status** Common non-breeding visitor to wetlands of C and W Trinidad, mostly Mar–Oct though some present year-round. Recently, a small party has visited SW Tobago each May–Sep.

**Least Tern**

adult breeding

adult non-breeding

immature

juvenile

**Brown Noddy**

adults

juveniles

immature

**Yellow-billed Tern**

adult non-breeding

juvenile

adult breeding

**Black Skimmer**

juvenile

adult breeding

adult breeding

adult non-breeding

## PIGEONS AND DOVES – COLUMBIDAE

A total of 309 species occur worldwide. Of these 15 have been recorded on Trinidad, eight also on Tobago. Ground doves favour open lowlands; quail-doves upland forest. The rest occupy a variety of habitats, and tend to be shy and wary on Trinidad, yet much more approachable on Tobago. Vocalisations can be collectively summarised as 'cooing', but each species has its own individual pitch, tone and rhythm.

### Scaled Pigeon *Patagioenas speciosa*                            30cm

Large and sleek. Adult male has reddish-brown head with red orbital. Underparts buff-white, with throat, breast and shoulders densely mottled brown and cream. Nape black, densely spotted white. Mantle, rump and wing-coverts reddish-brown with darker flight feathers and tail. Bill long, coral-red with white tip. Legs pinkish-grey. Female duller. Juvenile duller still, with indistinct dark scaling restricted to neck-sides, reddish-brown mottling to wing-coverts and lacks pale bill tip. **Voice** Low-pitched two-syllable *coo-oo*, audible over considerable distances; also a four-syllable *ooop cu coo coo* with emphasis on third note. **SS** Unlikely to be confused. **Status** Often perches prominently on exposed branches but very wary. Common resident of Trinidad's forests. Absent from Tobago.

### Scaly-naped Pigeon *Patagioenas squamosa*                        38cm

Very dark, large and bulky. Adult has dark grey head with reddish-orange skin around brown eyes. Underparts slate-grey. Nape and neck-sides maroon and distinctly scaled. Upperparts dark slate-grey, slightly paler on rump. Bill red, broadly tipped yellow; legs red. Juvenile overall more reddish-brown. **Voice** A four-note cooing with emphasis on last note. **SS** Unlikely to be confused. **Status** Accidental on Tobago: two birds in 2005.

### Band-tailed Pigeon *Patagioenas fasciata albilinea*                36cm

Full-breasted with long, broad tail. Adult male has vinous-grey head and red orbital around yellow eyes. Throat, breast and belly vinous-grey, paler on undertail-coverts. Nape bronze-green with conspicuous white hind-collar. Mantle and wing-coverts browner with dark grey rump and flight feathers. Tail two-toned grey, basally darker, distally paler, separated by black band. Bill and legs yellow. Female has greyer underparts and lacks green on nape. Juvenile much greyer and lacks collar. **Voice** Two-syllable *hu-whooo hu-whooo*. **SS** Unlikely to be confused. **Status** Extremely rare and localised resident of highest slopes in Trinidad's Northern Range, but not seen for many years and perhaps extirpated.

Scaled Pigeon

juvenile

♀

♂

Scaly-naped Pigeon

juvenile

♀

♂

Band-tailed Pigeon

juvenile

♂

♀

# PLATE 52: MID-SIZED PIGEONS

## Pale-vented Pigeon *Patagioenas cayennensis tobagensis* 30cm

Adult male has dark vinous forecrown with rest of head grey. Eyes and orbital bright red. Chin and throat grey-white. Breast and belly in front of legs vinous, rear belly, thighs and undertail-coverts white. Nape grey with metallic bronze gloss. Mantle reddish-brown, contrasting with grey rump and slightly darker grey-brown tail. Wing-coverts reddish-brown, flight feathers dark grey. Bill black, legs dark pink. Female duller. Immature recalls adult female. **Voice** A slow, low-pitched cooing, usually four notes with emphasis on second and fourth, *ke koo ke koo*. **SS** Unlikely to be confused. **Status** Locally common resident on both islands. Prefers open wet pasture and savannas in central and coastal lowlands of NE Trinidad; more widespread on Tobago.

## Eared Dove *Zenaida auriculata stenura* 23cm

Adult male has pinkish-grey crown. Face buff-pink with pale blue orbital ring and two narrow diagonal black lines on lower ear-coverts. Rest of underparts buff-pink with metallic sheen to lower neck-sides. Nape buff, mantle and rump olive-brown. Graduated tail grey-brown with all but central feathers broadly tipped rufous. Wing-coverts grey-buff with few large random black spots, flight feathers darker brown, fringed pale. Underwing-coverts white. Bill dark grey, legs coral-pink. Female and immature duller grey-brown. **Voice** A mournful three-note *ou-ee ou*. **SS** Unlikely to be confused. **Status** Common resident on both islands, most numerous in SW Tobago. Prefers open scrubby lowlands, mangrove edges, savannas and rice fields; also in lowland dry forest. Forms large flocks Aug–Sep.

## Feral Rock Pigeon *Columba livia* 30cm

The common feral pigeon of town and countryside. Highly variable in general coloration; often dark grey with purple and green neck sheen, black wing-bars, white rump and black tail-band. Other varieties are extensively white or reddish-brown. **Voice** A mournful cooing, often four notes with last one longer. **SS** Instant familiarity makes species unlikely to be confused. **Status** Abundant throughout country. Pigeon racing very common in Trinidad. Ubiquitous in habitat.

**Pale-vented Pigeon**

juvenile

adult

**Eared Dove**

♂

♂ juvenile

♀

♀ juvenile

♂ wild Rock Dove type

♀ typical dark feral city bird

**Feral Pigeon**

### Common Ground Dove *Columbina passerina albivitta*                16cm

Rather dainty. Adult male has pale grey crown with whitish face and throat. Rest of underparts pale pinkish-buff; breast, neck-sides and nape boldly scaled black. Upperparts pale grey-brown with random black spots on wing-coverts. Outertail black, rest grey-brown. Upperwing-coverts grey-brown tipped black, flight feathers rufous separated by buff carpal bar. Underwing-coverts rufous. Bill dainty, orange tipped black. Legs pink. Female lacks pink tones to breast and is browner above. Immature recalls female but has white upperparts flecking and less distinct black breast scaling. **Voice** A low-pitched and rather soft disyllabic *coo-oup*. **SS** From Ruddy and Plain-breasted Ground Doves by neck, breast and nape scaling and bill colour. Ruddy also lacks pale carpal bar. **Status** Locally common resident in low-lying, open dry areas of Trinidad, preferring agricultural land and savannas. Absent from Tobago.

### Plain-breasted Ground Dove *Columbina minuta minuta*                15cm

Very small, squat and short-tailed. Adult male has blue-grey crown and uniform pale pink face. Throat pale pink, breast and belly plain brownish-pink. Nape blue-grey, mantle grey-brown. Rump and tail blue-grey, outer tail feathers tipped black. Wing-coverts grey-brown with random large black spots. Flight feathers plain grey with reddish-brown inner webs and buff carpal bar. Underwing-coverts rufous. Bill greyish-yellow, legs pink. Adult female has duller underparts and browner upperparts. Immature recalls adult female, with buff fringes to upperparts. **Voice** A quiet single *cooo*. **SS** Male unlikely to be confused. Female from Ruddy Ground Dove by smaller size, brown rump and rufous underwing-coverts. From Common Ground Dove by lack of neck and breast scaling. **Status** Scarce Trinidad resident, favouring low-lying scrubby fields and savannas. Absent from Tobago.

### Ruddy Ground Dove *Columbina talpacoti rufipennis*                18cm

Adult male has powder blue-grey crown and nape. Face and throat buff with dark eyes and thin white orbital ring. Underparts dark pink. Mantle reddish-brown. Rump and central tail browner, with outer-tail feather tips black. Wings rufous, becoming orange on flight feathers, with scattered black tear-shaped spots on coverts. Underwing-coverts black. Bill pink, tipped darker. Legs pale pink. Adult female has greyer head, almost grey-white in some. Breast and belly grey, undertail-coverts white. Mantle and upperwing-coverts olive-brown, rump rufous. Immature recalls adult female. **Voice** Monotonous *hor hor hor* or *h h hor hor hor* with emphasis on final note. **SS** Adult male unmistakable. Female from Common Ground Dove by larger size, rufous rump, more rufous-winged in flight, lack of black neck and breast scaling, and black underwing-coverts. From Plain-breasted Ground Dove by much larger size, rump and underwing colour. From female Blue Ground Dove by smaller size and black wing-coverts spotting. **Status** An abundant widespread resident on both islands.

### Blue Ground Dove *Claravis pretiosa*                20cm

Adult male has blue-grey head, becoming blue-white on forecrown. Throat white, rest of underparts pale grey. Upperparts pale grey with black spotting on median coverts and black bar on greater coverts. Flight feathers dark grey. Tail grey with black outermost feathers. Bill dull yellow, legs pink. Female has reddish-brown head, buff-white throat and rest of underparts pale brown. Upperparts rich ruddy-brown with two chestnut bars on flight feathers. Central tail rufous, outer tail black. Immature recalls adult but has scaled mantle. **Voice** Single, well-spaced, hollow notes, *horp...horp...horp*. **SS** Adult male unmistakable. Female from all other ground doves by larger size and red-spotted wing-coverts. **Status** A rare, shy Trinidad resident favouring lowland forest and savanna. Absent from Tobago.

### Scaled Dove *Columbina squammata ridgwayi*                22cm

Sleek, long-tailed and short-legged. Adult has buff-white forecrown; crown and face buff-brown densely scaled black. Underparts buff-white, heavily scaled black. Upperparts buff-brown scaled black. Flight feathers grey-brown with rufous fringes conspicuous in flight. Four central tail feathers brown, next four black tipped white, outermost two all white. Bill black, legs pale pink. Immature recalls adult. **Voice** A monotonous three notes, first drawn-out, *aau cohon*. **SS** Unlikely to be confused. **Status** Accidental on Trinidad; two birds in 1926. No records from Tobago.

## Plain-breasted Ground Dove

## Common Ground Dove

juvenile

♀

♂

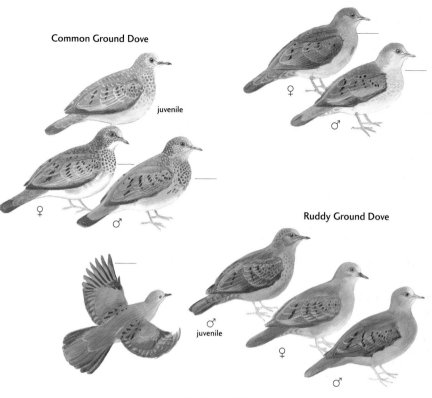

♀

♂

## Ruddy Ground Dove

♂
juvenile

♀

♂

## Blue Ground Dove

juvenile

♀

♂

## Scaled Dove

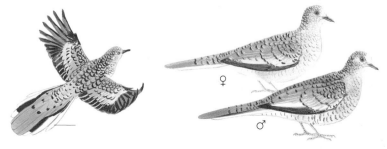

♀

♂

# PLATE 54: LARGER DOVES

### Eurasian Collared Dove *Streptopelia decaocto* 33cm

Adult has head and underparts pale grey. Nape pale grey with white-bordered black hindneck collar. Mantle, inner wing-coverts and rump pale brownish-grey. Outer wing-coverts whitish contrasting with dusky flight feathers. Underwing-coverts pale grey, flight feathers darker. Rump, central four tail feathers and basal half of rest pale grey-brown, distal half of outer feathers white. Undertail mainly white. Bill black, legs dull pink. Immature browner and lacks collar. **Voice** A three-syllable *coo* with second drawn-out. **SS**. Unmistakable. **Status**. In recent years, several documented records for Trinidad (presumed escaped cagebirds, but species is spreading through West Indies).

### White-tipped Dove *Leptotila verreauxi* 28cm

Adult on Trinidad (*L. v. zapluta*) has pinkish-grey head with turquoise-blue orbital ring. Throat and upper breast warm pinkish-grey, paler on lower breast and grey-white on belly and undertail-coverts. Upperparts olive-brown with bronzy-copper nape patch. Central rectrices brown, rest black, with white tips to outer four diminishing in size towards centre, showing as opposing white wedges on take-off. Underwing-coverts chestnut. Bill black, legs coral-pink. Adult on Tobago (*L. v. tobagensis*) similar, but is bronze-green on nape, lacking coppery reflections, throat is more extensively and purer white, underparts paler and white ventral area more extensive. Young like browner adults, with pale fringes to upperparts affording a scaled effect, and has forehead concolorous with crown. **Voice** A quiet, mournful, drawn-out disyllabic *ooo-u*. **SS** From Grey-fronted Dove by concolorous crown, turquoise orbital ring, general facial colour and voice. On Trinidad, both are very timid and many views are of birds flying off; note greater amount of black in tail of Grey-fronted with white restricted to outer two tips. **Status** Widespread, common resident in T&T, mainly at lower elevations. A ground-dweller, its shy nature belies its abundance on Trinidad; on Tobago much more approachable.

### Grey-fronted Dove *Leptotila rufaxilla hellmayri* 28cm

Adult has powder-blue crown, almost blue-white on forecrown. Face and throat tawny-buff with yellow eyes and narrow red orbital ring. Upper breast vinous-buff, becoming grey-white on belly and undertail-coverts. Nape grey with vinous wash. Upperparts olive-brown with darker flight feathers. Tail basally olive-brown, distally black with narrow white tips on outermost two rectrices. Bill black, legs dull red. Juvenile recalls adult, but paler and browner, and scaly. **Voice** Similar in tone to White-tipped Dove but monosyllabic, a drawn-out *oooo*. Increases in resonance halfway through, before falling again. **SS** From White-tipped Dove by blue-white forecrown, brighter face, red orbital ring and voice. **Status** Uncommon widespread resident on forest floor on Trinidad, being more common above 200m. Absent from Tobago.

### Lined Quail-Dove *Geotrygon linearis trinitatis* 28cm

Adult male has rich brown crown and buff forehead. Face buff with short black loral stripe, long black moustachial, white border to rear ear-coverts and reddish-brown orbital. Throat white; rest of underparts dull buff-brown. Upperparts deep russet-brown with purple sheen and scaly feather fringes to upper mantle. Bill black, legs dull pink. Female duller. Juvenile much browner, heavily scaled above and barred on breast and sides. **Voice** Low-pitched, rather drawn-out *hoouu..hoouu*. **SS** From Ruddy Quail-Dove by larger size, bolder face pattern and darker brown upperparts. **Status** Rare resident restricted to forested tops of Northern Range, probably not below 900m. Historically recorded from Main Ridge forest in Tobago, but not seen there for many years.

### Ruddy Quail-Dove *Geotrygon montana montana* 23cm

Adult male has reddish-brown crown and upper face. Broad cream stripe runs over base of cheeks and ear-coverts, below which is dark red malar stripe; also cream bar above base of bill and thin red orbital. Chin and throat cream, breast and belly dull reddish-brown. Upperparts and underwing bright reddish-brown, with purple gloss to neck-sides and cream line at wing bend. Bill dark red, legs brighter coral-red. Adult female has duller face pattern, browner underparts and brownish-olive upperparts. Juvenile browner and paler. **Voice** Long, drawn-out, deep *ooou*. **SS** From Lined Quail-Dove by smaller size, more uniform paler coloration, face markings and dark red bill. **Status** Very scarce, shy resident in forested Northern Range of Trinidad. No documented records for Tobago.

**Eurasian Collared Dove**

**White-tipped Dove**

juvenile

*tobagensis*
(Tobago)

*zapluta*
(Trinidad)

juvenile

**Grey-fronted Dove**

♂

♀

**Lined Quail-Dove**

juvenile

♂

♀

**Ruddy Quail-Dove**

juvenile

♀

♂

## MACAWS AND PARROTS – PSITTACIDAE

This family comprises 342 species found principally in tropical and subtropical areas of Asia, Australasia, Africa and the Neotropics. Ten species have been recorded on Trinidad, of which six are regular, and two also occur on Tobago. Macaws are medium-sized to large with bare facial skin, long wings and long tails. Parakeets are similar in size and shape to small macaws, but lack bare facial skin. Parrots are quite large and chunky with broad, rounded wings, and parrotlets are small, dumpy, short-winged and short-tailed.

### Blue-and-yellow Macaw *Ara ararauna*                                     84cm

Adult has bright blue crown with apple-green forehead. Facial skin white with narrow black lines and lemon-yellow eyes. Throat black, rest of underparts bright golden-yellow. Undertail-coverts blue, undertail lemon-yellow. Upperparts azure-blue, flight feathers darker royal-blue. Underwing lemon-yellow. Bill and legs black. Immature recalls adult. **Voice** Variety of shrieking cries, including a raucous *kraah* and *raak* in flight. **SS** Unmistakable. **Status** Locally extirpated in 1970. A reintroduction programme is underway in woodland bordering extensive swamp at Nariva, E Trinidad. Absent from Tobago.

### Scarlet Macaw *Ara macao macao*                                     91cm

Adult has red crown and unmarked chalk-white facial skin around pale yellow eyes. Underparts all red, except blue undertail-coverts which contrast with red undertail. Nape and mantle red, rump and uppertail-coverts blue. Tail red, tipped blue. Scapulars and lesser coverts red, median and greater coverts yellow, tipped green; flight feathers dark blue. Large chunky bill white above, dirty black below. Legs dark grey. Immature recalls adult, but has brown eyes. **Voice** A deep guttural and grating *raahk*. **SS** Possible confusion when perched with Red-and-green Macaw *A. chloropterus* (unrecorded in wild in Trinidad, but escapees occur), but separated by unlined facial skin and yellow wing-coverts. **Status** Two sightings on Trinidad, the most recent at Waller Field in 1943. Absent from Tobago.

### Red-bellied Macaw *Orthopsittaca manilata*                                     51cm

Adult has greenish-blue crown, bluer on forehead. Facial skin bright, pale yellow. Underparts dull grey-green with red patch on lower belly. Upperparts green, wings blue with greenish coverts. Undertail and underwings greenish-yellow. Bill and legs black. Female has less blue on head, whilst juvenile and immature recall adult but have white facial skin and pale horn-coloured bill tip. **Voice** A high-pitched, shrill series of shrieks and squawks, including a sharp *kareek*. **SS** Unlikely to be confused. **Status** Very local but fairly common resident on Trinidad, found mainly in Nariva swamp and the central savannas where Moriche Palms occur. Absent from Tobago.

### Brown-throated Parakeet *Aratinga pertinax surinama*                                     23cm

Adult has pale blue crown with grey forehead. Face orange, browner on ear-coverts. Throat and upper breast brownish-green, more yellow-green on lower breast and belly. Upperparts varying shades of green, flight feathers blue. Underwing greyish-yellow. Bill dark grey-brown, legs blackish. Immature like adult. **Voice** Flight-call a guttural *jakk*. **SS** Size and shape render it unlikely to be confused. **Status** Rare visitor to lowlands of C & W Trinidad. Absent from Tobago.

Blue-and-yellow Macaw

Scarlet Macaw

Red-bellied Macaw

♂

♀

juvenile

Brown-throated Parakeet

### Yellow-crowned Amazon *Amazona ochrocephala ochrocephala* 36cm

Stiff, shallow wingbeats. Adult bright yellow from forehead to crown (increasing with age), rest of head emerald-green with broad white orbital ring. Underparts slightly paler green. Upperparts dark green. Tail brighter lime-green, yellower on distal half with obscure red patch at base of outer feathers. Wings bright lime-green with black primary tips, royal blue secondary tips, bright red square speculum on secondaries and inconspicuous red patch at wing bend (usually only seen in flight). Bill pale grey, tipped black; legs grey, eyes yellow-orange with red orbital. Juvenile and immature have small, variable yellow patch on forecrown that increases with age; eyes brown. **Voice** A metallic *kii-ao* or *kirrao*. **SS** From Orange-winged by brighter green plumage, lack of blue on head, orbital ring, red wing bend and red speculum. **Status** Uncertain in Trinidad. Birds, perhaps of captive origin, now regular in suburban areas along east–west corridor. Also forested edges of lowland marshes and swamps E and SW Trinidad. Breeding suspected but unproven. Absent from Tobago.

### Orange-winged Amazon *Amazona amazonica* 33cm

Stiff, shallow wingbeats. Adult has pastel green head with extremely variable amounts of blue and yellow that increase in size and intensity with age. Body feathers pastel green. Wings brighter with square orange speculum just visible in folded wing and black primary tips. Tail bright green with yellow-green outer feather tips. Bill pale horn with darker tip; legs pale grey; eyes yellow to deep orange. Juvenile duller with orange speculum and significantly less blue and yellow on head; eyes brown. **Voice** Varied shrieks and screams, most frequently metallic couplets, *qwek qwek*. **SS** From larger Yellow-crowned by face pattern, no orbital, and orange speculum. **Status** Abundant, widespread resident on Trinidad. Very common on Tobago, but prefers hill forests.

### Blue-headed Parrot *Pionus menstruus menstruus* 25cm

Adult has entire head deep blue with pink flecks on chin. Lower breast to vent turquoise-green, undertail-coverts blood red. Golden wash to wing-coverts and outer rectrices blue. Juvenile has powdery blue head and red scallops on undertail-coverts. **Voice** A harsh two-syllable *kee wik* in flight. **SS** From *Amazona* parrots by smaller, more compact shape, solid blue head and breast, and more deliberate, deep wingbeats. **Status** Uncommon resident of hill forest and central savannas on Trinidad. Absent from Tobago.

### Green-rumped Parrotlet *Forpus passerinus viridissimus* 14cm

Grass green from forehead to back, and tail. Face to undertail-coverts, rump/uppertail-coverts brighter, paler green, with turquoise-blue wash to outer edge of wing. Bill and legs ivory. Female lacks blue on wing and pale yellowish forehead; juvenile similar. **Voice** A high-pitched, chattering *twit-it twit-it twit-it*. **SS** Unmistakable. **Status** Common resident lowland T&T. Prefers savannas, mangroves and freshwater swamps.

### Scarlet-shouldered Parrotlet *Touit huetii* 18cm

Mostly deep green, paler below; yellowish undertail-coverts, ochraceous wash to head, and forehead and foreface rich blue-black. Wing-bend and underwing-coverts scarlet, outer wing-coverts royal blue, primaries dark brown with pale green leading edge. Central tail green, rest deep red with blackish tip. Broad white orbital ring, dark brown eyes, ivory bill and flesh-coloured legs. Female less extensively blue on face/wings, and lacks red in tail; young lack any blue on face. **Voice** A high-pitched, shrill, frenetic chatter, *weerch weerch weerch*; similar to, but each note longer than, Lilac-tailed. **SS** Unlikely to be confused. **Status** Accidental Trinidad with no documented records since 1980.

### Lilac-tailed Parrotlet *Touit batavicus* 15cm

Most frequently seen in noisy, tight, high-flying groups. Adult has green crown and ear-coverts, more yellow on lores and lower cheeks. Neck-sides and upper breast blue-green, faintly scalloped. Rest of underparts yellow-green. Mantle and rump black. Tail pale lilac, tipped black. Intricate wings markings: shoulder black, greater coverts lime-yellow forming broad bar, with each feather tipped blue, tertials lime-yellow, flight feathers black, and obscure leading edge of underwing red. Bill yellowish-grey, legs dull orange. Young recalls dull adult. **Voice** Flight-call a shrill, nasal scream, *scree-eet*. Usually quiet when feeding. **SS** Unmistakable. **Status** Uncommon but widespread resident on Trinidad, in forest, suburban parks, estates and lightly wooded savannas. Absent Tobago.

**Yellow-crowned Amazon**

juvenile

adult

immature

**Orange-winged Amazon**

juvenile

adult
(variant)

adult

**Blue-headed Parrot**

juvenile

adult

**Green-rumped Parrotlet**

♀

♂

**Scarlet-shouldered Parrotlet**

juvenile

♀

♂

**Lilac-tailed Parrotlet**

juvenile

adult

# CUCKOOS – CUCULIDAE

A family of 141 species worldwide. Nine species occur on Trinidad, just three on Tobago. Variable in overall size, they are generally slender with long tails and slightly decurved bills. Most are shy, but anis are distinctly social. Striped Cuckoo is the only brood-parasite found in our region.

### Striped Cuckoo *Tapera naevia naevia* 27cm

Adult has black crown with rich chestnut tips to feathers and an often erectile crest, especially when calling. Face grey-buff with creamy white supercilium above eye and sweeping down neck-sides; smudgy, broad, dark brown postocular line forms lower border to supercilium, and indistinct black malar. Underparts creamy white with buff undertail-coverts. Upperparts brown with black and buff stripes. Tail long, graduated and plain grey-buff. Wings short with dull grey-brown flight feathers. Bill yellowish-grey, legs grey. Juvenile similar, with rufous and buff underparts spotting and faint upper-breast barring. **Voice** Two-note musical whistle, the second note higher and repeated at regular intervals, often from an exposed perch. Also, a rich, ascending *dwee dwee dwee dwee-eu*. **SS** Unmistakable. **Status** Uncommon resident of open lowland pastures, mangrove edges, scrubby hillsides and savannas on Trinidad. Absent from Tobago.

### Black-billed Cuckoo *Coccyzus erythrophthalmus* 30cm

Adult has olive-grey head with very narrow red orbital ring. Underparts silky white; undertail grey with small inconspicuous white tips. Upperparts olive-brown with pale wing-coverts and scapular tips. Long, graduated tail is olive-brown with small white tips. Bill black, slight in structure; legs dark grey. Immature duller, less clean-cut, with yellow orbital, pale fringes to most coverts and scapulars, and lacks spotted undertail. **Voice** Usually silent. **SS** From Yellow-billed Cuckoo by upperparts and undertail colour, and lacks rufous on wings. **Status** Accidental on Trinidad, with just one Sep record. Favours open scrub and lowland forest.

### Yellow-billed Cuckoo *Coccyzus americanus* 30cm

Adult has soft grey-brown head with black mask through eyes. Underparts silky white. Upperparts grey with rufous panel (duller rusty in some) to folded wing. Graduated tail with central feathers grey and rest black with large terminal white spots. In flight, flight feathers appear rufous with black tips; underwing-coverts white. Bill dark horn above and yellow below. Legs dark grey. Immature recalls adult, with yellow orbital ring and diffuse white undertail spots. **Voice** Usually silent. **SS** From all similar cuckoos by rufous wing-panel. **Status** Rare migrant to T&T, preferring dry lowland savanna. Most records Oct–Nov, but has overwintered.

### Mangrove Cuckoo *Coccyzus minor* 32cm

Adult has grey head with broad black mask from bill through eyes to rear ear-coverts, and yellow orbital. Chin and throat grey-white, breast and belly pale buff (pale morph) or rich cinnamon, including underwing-coverts (dark morph). Upperparts grey-brown with blacker flight feathers. Graduated black tail has large white spots on outer feathers. Bill stout; black above, bright yellow below. Legs dark grey. Immature recalls adult, with subdued mask and tail markings. **Voice** Harsh, sharp and nasal *ki ki ki* or *arn arn*. **SS** From Dark-billed Cuckoo by larger size and bill colour. From Yellow-billed Cuckoo by underparts colour and lack of rufous in wings. **Status** Rare resident in T&T. Prefers mangroves of W Trinidad and SW Tobago.

### Dark-billed Cuckoo *Coccyzus melacoryphus* 25cm

Adult has dark grey-brown crown and upper face, with black mask from lores to ear-coverts, and narrow yellow orbital ring. Lower face, throat and neck-sides grey, rest of underparts rich buff. Upperparts dark olive-brown. Graduated tail black with broad white spots on undertail. Bill black, legs grey. Immature duller, with brown crown and grey tail tip. **Voice** Series of harsh guttural notes. **SS** From Mangrove Cuckoo by smaller size, face, neck and bill colour. **Status** Very rare visitor to mangrove and lowland scrub on Trinidad, where has bred. Absent from Tobago.

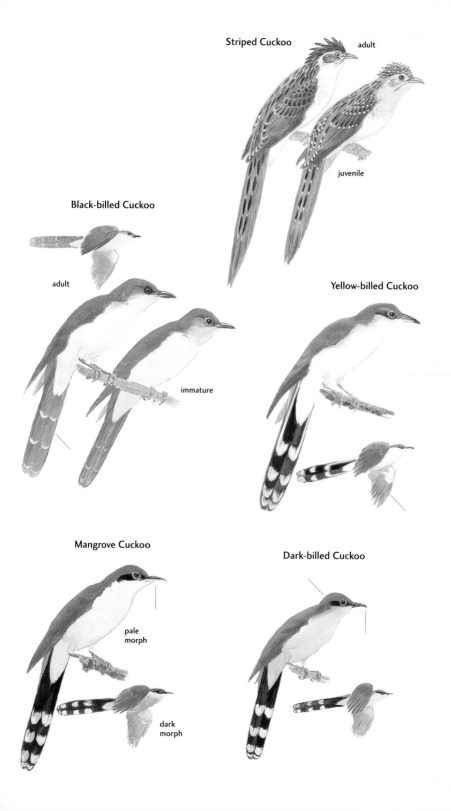

Striped Cuckoo

adult

juvenile

Black-billed Cuckoo

adult

immature

Yellow-billed Cuckoo

Mangrove Cuckoo

pale
morph

dark
morph

Dark-billed Cuckoo

### Squirrel Cuckoo *Piaya cayana insulana* 43cm

Large and very long-tailed. Adult has reddish-orange head with pale green orbital ring around red eyes. Throat and upper breast buff, lower breast and belly grey. Undertail-coverts dark grey, undertail dark brown with large white spots. Upperparts chestnut-brown with broad black subterminal band and white tips to tail feathers. Bill robust and greenish-yellow, legs grey. Immature recalls pale adult, with less white on tail tip. **Voice** Varied sharp calls, most commonly a harsh two-syllable *ki-harr*, chattering *rik-ki-ki-yaa* and sharp *stk*. **SS** From Little Cuckoo by much larger size, longer tail and underparts colour. **Status** Uncommon widespread resident in Trinidad's forests. Absent from Tobago.

### Little Cuckoo *Piaya minuta minuta* 25cm

Small, short-winged and long-tailed. Adult has cinnamon-brown head with thin red orbital ring. Throat and breast warm chestnut, belly and undertail-coverts grey. Undertail black with broad white bars. Upperparts cinnamon-brown. Graduated tail darker brown, tipped white. Bill short and yellow, legs grey. Immature much darker brown with black bill and lacks tail spots. **Voice** Series of harsh single notes, *tek* or *kik*, and a metallic, rather hollow series of 3–4 rising notes followed by 8–10 on same pitch; also quiet nasal *kuk*, followed by an even quieter but harsh *yarh*. **SS** From Squirrel Cuckoo by smaller size and underparts colour. **Status** Scarce and secretive Trinidad resident, favouring low-lying open savannas, second-growth scrub and mangrove edges. Absent from Tobago.

### Greater Ani *Crotophaga major* 43cm

Long-tailed and huge-billed with shaggy nape and conspicuous yellow-white eyes. Adult entirely glossy black, but wings and tail have blue or green sheen, according to light. Bill black with pronounced ridge on upper mandible. Legs black. Juvenile recalls adult, but has smaller culmen ridge and dark eyes. **Voice** Series of raucous, coughs and croaks, *grrrkh* or *arrkh*, plus cacophony of bubbles and squawks. **SS** From Smooth-billed Ani by much larger size, longer tail, glossy plumage and eye colour. **Status** Locally uncommon resident of mangrove swamps and adjacent scrub on Trinidad. Usually shyer than Smooth-billed Ani. Absent from Tobago.

### Smooth-billed Ani *Crotophaga ani* 33cm

Short, rounded wings and long tail. Adult all black with pale fringes to upper breast and head. Nape feathers broadly edged pale grey and ruffled, giving shaggy look. Wing and tail feathers black with glossy fringes to coverts. Bill large and cone-shaped with raised flat ridge on basal half of upper mandible; from black through lead grey to grey-buff. Legs black. Immature smaller and less glossy. **Voice** Plaintive, two syllables with second higher pitched, *ooer-eek!* Also varied cackles and grunts. **SS** From Greater Ani by smaller size, shorter tail, matt black plumage and dark eyes. **Status** Very common and widespread resident in T&T.

Squirrel Cuckoo

Little Cuckoo

Greater Ani

Smooth-billed Ani

## BARN OWLS – TYTONIDAE

Family of 16 species worldwide. Characterised by heart-shaped facial disc and long legs, just one species is found in T&T.

### Barn Owl *Tyto alba hellmayri*     38cm

Large and ghostly pale with dark eyes and no ear-tufts. Adult has warm buff crown mottled black and white. Facial disc white. Most have entire underparts white with sparse dark chevrons, but scarce tawny morph has breast and belly warm tawny-buff. Upperparts warm buff, almost ginger, heavily mottled blue-grey and speckled black and white. Underwing white. Bill pale yellowish-pink, legs feathered white. Immature recalls adult. **Voice** Varied hisses and shrieks. **SS** Unmistakable. **Status** Uncommon resident in T&T. Nocturnal and crepuscular, favouring open scrubby areas, grasslands and water meadows.

## OWLS – STRIGIDAE

Some 213 species occur worldwide of which seven species have occurred in T&T. Only three are regular on Trinidad, whilst a fourth is a secretive and rare resident and one is resident on Tobago; the other two are vagrants. Highly variable in size; mainly nocturnal, shy and wary by nature. Most are cryptically plumaged in shades of brown, with piercing yellow or orange eyes.

### Tropical Screech Owl *Megascops choliba crucigerus*     23cm

Four morphs, with intergrades. The common morph adult has grey-brown crown and short ear-tufts densely streaked black. Facial disc buff-brown, bordered black, with white eyebrows and yellow eyes. Breast and belly grey-buff with fine white vermiculations overlaid by sparse bold black streaks. Upperparts grey-brown streaked and spotted black and white. Scapulars have prominent row of white spots. Flight feathers darker brown, barred black. Bill pale grey, feet brown. There are rarer rufous, deep rufous and dark grey-brown phases. Juvenile morphs as adult, but barred. **Voice** Series of 4–6 tremulous low-pitched hoots followed by a louder, slightly higher hoot, *prrrr prrrr prrrr prrrr per.* **SS** Presence of ear-tufts combined with small size eliminates all other owls. **Status** Uncommon but widespread throughout Trinidad, but infrequently seen due to nocturnal habits. Prefers suburban gardens, cultivated estates, open scrub and light forest, usually at lower elevations. Absent from Tobago.

### Spectacled Owl *Pulsatrix perspicillata trinitatis*     46cm

Adult has chocolate-brown crown and facial disc, with broad white eyebrows that meet above bill and become moustachial stripes below eyes, affording 'spectacled' effect. Eyes yellow. Throat band white, contrasting with chocolate-brown upper breast. Rest of underparts warm ochre-buff. Upperparts chocolate-brown with faint darker barring to flight and tail feathers. Bill dirty yellowish-horn, feet grey. Juvenile has entire body plumage initially creamy-white, with greater wing-coverts barred brown, blacker flight feathers and black, heart-shaped facial disc. Upperparts gradually become patchily brown with white tail tip. **Voice** Series of deep bubbly *whoop* notes, with last few lower pitched. Also 5–6 quick, low-pitched hollow notes, *buk buk buk buk buk.* **SS** Unmistakable. **Status** Scarce resident of cultivated estates and forests on Trinidad, occasionally wooded suburban areas. Absent from Tobago.

## Barn Owl

## Tropical Screech Owl

deep rufous

rufous

light brown

dark grey-brown

## Spectacled Owl

adult

juvenile

### Ferruginous Pygmy Owl *Glaucidium brasilianum phalaenoides*                    15cm

Adult has brown crown, densely streaked white. Facial disc buff, streaked grey with narrow white eyebrows and yellow eyes. Throat, breast and belly white, boldly streaked reddish-brown; undertail-coverts white. Nape brown with black-and-white chevron (the 'false eyes'), but often difficult to see. Mantle and rump brown. Wing-coverts brown, heavily spotted white; flight feathers brown, barred buff-white. Tail reddish-brown with indistinct darker bars. Bill yellowish-grey, feet bright yellow. Some birds more rufous. Juvenile recalls adult but more spotted than streaked. **Voice** Series of short *hoop* notes on even pitch and repeated 10–30 times. Often calls during day. **SS** From Tropical Screech Owl by smaller size, lack of ear-tufts, less prominent white crown streaking, greater contrast on underparts and nape chevrons. **Status** Common, widespread, diurnal Trinidad resident. Several undocumented reports from Tobago.

### Burrowing Owl *Athene cunicularia brachyptera*                    24cm

Short-bodied with long legs, short tail and no ear-tufts. Adult has dark brown crown densely spotted white. Facial disc grey with broad, conspicuous white eyebrows that meet above bill; diffuse white crescents below eyes and broader white crescents bordering cheeks. Eyes golden-yellow. Ear-coverts, throat and neck-sides brown, mottled white, bordered by full white nuchal collar. Upper breast dark brown with large white spots; lower breast and belly dirty cream-white, barred brown. Upperparts dark brown spotted white. Flight feathers and tail brown barred buff. Bill pale horn, legs feathered creamy white. Juvenile recalls adult. **Voice** Fluty *kar kar* alarm. **SS** Unlikely to be confused. **Status** Accidental to SW Trinidad where bred in 1992. Diurnal, found in open country.

### Mottled Owl *Ciccaba virgata virgata*                    36cm

Medium-sized, rotund and large-headed with no ear-tufts. Adult has dark grey-brown crown. Facial disc brown, with two narrow buffy eyebrows and black eyes. Throat and upper breast dark grey-brown densely streaked black, with narrow band of warm buff down centre to lower breast, widening on belly and flanks, with sparse black streaking. Upperparts dark brown finely mottled buff, with row of buff oval spots on scapulars. Flight feathers dark grey-brown heavily spotted black and buff. Tail fairly long, brownish-black with several broad pale grey or buff bands. Bill yellowish-grey, feet brown. A very rare pale morph. Juvenile has body plumage initially buff, a pale facial disc and darker grey upperparts barring. **Voice** A long three-syllable note, the middle one higher pitched, *hu ow oo*; a three-note *houw houw houw*; and a sharp, nasal *reear*. **SS** Combination of size, no ear-tufts and black eyes make it unlikely to be confused. **Status** Rare nocturnal resident of Trinidad's forests. Absent from Tobago.

### Striped Owl *Asio clamator oberi*                    36cm

Medium-sized, pallid and boldly streaked, with prominent erect ear-tufts. Adult has rich buff crown densely streaked black, buff-white facial disc bordered black, with dark orange eyes and bold white eyebrows. Underparts buff-brown, boldly striped black. Upperparts rich buff-brown, boldly streaked black. Flight feathers and tail brown, barred dusky. Bill and feet black. Juvenile has striping replaced by duller barring. **Voice** Seldom heard: series of up to six muffled, deep barks, *owr owr owr*. **SS** See Short-eared Owl. **Status** Rare resident on Tobago; prefers open scrub and forested edge. Strictly nocturnal. A day-roosting bird found 10m high in trees in dense forest. Several undocumented reports from open grasslands on Trinidad.

### Short-eared Owl *Asio flammeus pallidicaudus*                    35cm

Medium-sized, bulky with inconspicuous ear-tufts. Hunts low over open ground; often hovers, quartering like a harrier. Adult has milk chocolate-brown crown streaked buff. Facial disc warm buff, becoming creamy near bill; disc of blackish feathers surrounds each yellow eye. Ear-tufts only protrude just above crown when head in profile. Throat and upper breast warm buff densely streaked dark brown. Lower breast creamy buff becoming creamy white on belly with very sparse, narrow black streaks. Upperparts chocolate-brown spotted buff and white. Wing-coverts basally dark brown, distally buff with black patch at wing bend; cream leading edge to forewing and golden-buff flight feathers distally spotted black. Underwing very pale with several rows of black spots on outer primaries and black carpal 'comma'. Bill and legs grey. Juvenile recalls adult. **Voice** Usually silent. **SS** From Striped Owl by bulkier jizz, lack of prominent ear-tufts, yellow eyes, lack of rufous tones to upperparts, buff facial disc and lack of dense underparts streaking. **Status** Partially diurnal. Accidental on Trinidad, with one record of two at the Caroni Rice project, Sep 2001.

## Ferruginous Pygmy Owl

grey
morph

brown
morph

rufous
morph

## Burrowing Owl

## Mottled Owl

## Striped Owl

## Short-eared Owl

## OILBIRD – STEATORNITHIDAE

Just one species of oilbird, the only nocturnal fruit-eating bird in world. They roost by day in caves and grottos, emerging at nightfall with much clicking and fluttering.

### Oilbird *Steatornis caripensis* 46cm

Large, long-winged and long-tailed with large hooked bill and short legs. Male entirely dark rufous-brown with indistinct white spotting on crown and underparts; much bolder spotting on wing-coverts, flight and outer tail feathers. Female paler and more rufescent. **Voice** Screeches, moans, clucks and a guttural snarl. **SS** Unmistakable at rest. From all other 'nightbirds' by much larger size and uniform colour. **Status** Localised Trinidad resident. Strictly nocturnal. Roosts in deep caves using echolocation to navigate. May fly *c.*100km in search of fruiting palms. A record of several feeding at Hillsborough dam, Tobago, Sep 1988.

## POTOOS – NYCTIBIIDAE

Potoos are nocturnal birds with melancholic wails associated with death and disaster in local mythology. They have remarkably cryptic plumage and it is easy to pass a roosting bird as it perches upright on a tree stump or broken branch, in plain view.

### Common Potoo *Nyctibius griseus griseus* 38cm

Grey-brown head densely streaked black, with orange or orange-yellow eyes and darker malar stripe. Underparts grey with black breast spotting and variable black mottling and streaking. Upperparts grey-brown finely streaked black and fringed buff; tail grey with irregular black and white bands. Scapulars broadly tipped white, wing-coverts grey-brown fringed buff, black or white; flight feathers dark grey with pale barring. Bill feathered, legs brown. Juvenile smaller and paler. **Voice** Series of four slow, mournful descending whistles. **SS** Day-roosting bird unlikely to be confused, except with a branch! After dark, from other 'nightbirds' by large size, long tail and orange eyes. **Status** Locally uncommon but widespread resident in swamps and forests of T&T. At night, hunts from regular perch, usually within 1m of ground.

## NIGHTJARS AND NIGHTHAWKS – CAPRIMULGIDAE

Nightjars are strictly nocturnal and prefer to hunt from a perch. Nighthawks are also crepuscular and hunt more exclusively on the wing. Both are cryptic in colour; sexes often differ in position or presence of white on flight and tail feathers. On Trinidad, four species are resident and two migrant visitors. Just one resident nightjar and one migrant nighthawk on Tobago with a record of an unidentified larger nighthawk, Dec 2001.

### Rufous Nightjar *Caprimulgus rufus minimus* 25cm

Large and round-winged, with white in tail but not wing. Adult male has reddish-brown head heavily barred and streaked black. Underparts reddish-brown densely vermiculated black, with inconspicuous pale buff band on upper breast. Upperparts reddish-brown, heavily streaked or barred black, with scapulars and wing-coverts fringed and tipped buff. Tail browner with distal half of outer four feathers white. Female paler and browner with no white in tail. Juvenile has whiter crown, more pronounced breast barring and white scapular markings. **Voice** *Chup wup wup weeuu*. **SS** From Pauraque by face and throat colour, lack of white or buff flight-feather markings and restricted white in tail. **Status** Scarce and rarely seen Trinidad resident, favours dry woodland. Rarely sits on roads. Most frequently encountered on Bocas Is or higher-elevation forest on Mt St Benedict. Absent from Tobago.

### White-tailed Nightjar *Caprimulgus cayennensis leopetes* 22cm

Small and delicate in structure. Adult male has dark brown head with buff supercilium. Chin pale buffy-grey, rest of underparts white with faint but dense buff and black chevrons. Undertail white. Nape grey-brown with obvious rufous hind-collar. Upperparts dark brown mottled black, with scapulars fringed buff. Wings dark grey-brown with coverts tipped white and white bar on outer primaries. Central tail feathers grey-brown mottled black, outer three white with brown line. Female has buff underparts, buff spotting on outer flight feathers and lacks white in tail. Immature recalls adult female. **Voice** Very high-pitched almost hissing whistle. One short note followed by long drawn-out second, *huu heeeeeu*, or slightly higher *zup weeeeu*. **SS** Day-roosting birds easily identified by small size and chestnut collar. At night, from Pauraque by smaller size, paler coloration and shorter tail. **Status** Common resident on both islands, favouring open scrubby areas, and sits on roads and tracks at night. Roosts in dry woodland.

Oilbird

Common Potoo

White-tailed Nightjar

Rufous Nightjar

♂

♀

♂

♀

### Pauraque *Nyctidromus albicollis albicollis*                                    28cm

At rest, tail projects well beyond wing-tip. Adult male has dark grey crown lightly streaked black. Face pattern distinctive, with warm ruddy cheeks and thin white moustachial. Throat crescent white, breast and belly rich brown heavily barred black. Upperparts dark grey mottled black, with wing-coverts and scapulars heavily spotted white or buff. Flight feathers black with band of bold white spots at base of primaries. Outermost two tail feathers black, next two white, rest black; all narrowly barred buff. Rare rufous morph. Female differs in wing and tail markings: spotting at base of primaries is buff and white in tail is restricted to tip. Juvenile recalls adult with paler throat and subdued white or buff wing and tail markings. **Voice** Full-toned whistle rising on middle syllable, *per weee weoo*. **SS** From White-tailed Nightjar, which often shares habitat, by larger size, darker overall colour and longer tail. From Rufous Nightjar by bold flight-feather spotting and more prominent tail markings. **Status** Common Trinidad resident, favouring savanna scrub, lowland forest and mangrove edges. Often seen on forest roads and tracks at night. Absent from Tobago.

### Short-tailed Nighthawk *Lurocalis semitorquatus semitorquatus*                  22cm

Almost strictly nocturnal, seen at last light and in pre-dawn period. Dashing flight usually just above treetop level, like a large, long-winged bat. At rest, wings protrude well beyond tail. Adult male appears all black with inconspicuous white throat crescent, pale patch on scapulars and tertials, and small white tail tips. At daytime roost in canopy, rufous barring to lower breast and belly conspicuous. Female has even less noticeable pale feathering. Juvenile recalls adult. **Voice** A quiet hissing *whiss whiss*. **SS** From all 'nightbirds' by long pointed wings and short square tail without obvious white wing markings. White in tail restricted to tip. **Status** Locally scarce but widespread resident on Trinidad, favouring forest edge. Absent from Tobago.

### Lesser Nighthawk *Chordeiles acutipennis acutipennis*                           20cm

At rest, wing-tips reach tail tip. Adult male has dark grey-brown head. Chin and throat white, rest of underparts grey densely barred dark brown and rufous. Upperparts grey-brown with narrow bands of dark grey, black, buff and white. Tail slightly notched, banded black and grey with white subterminal band on underside. Flight feathers black with broad white stripe across centre. Bill black. Female has buff throat and flight-feather band, and lacks white in tail. Juvenile recalls adult. **Voice** Fast drawn-out hollow trill. **SS** From both Common *C. minor* and Antillean Nighthawks *C. gundlachii* (no acceptable records in T&T), with extreme care, by smaller size, more rounded wing-tips and white wing-stripe nearer wing-tip. **Status** Uncommon Trinidad resident; numbers augmented in Jun–Oct by visitors from mainland. Prefers mangrove-lined west coast. One over Bon Accord, Dec 2006, was first documented record for over 50 years.

### Nacunda Nighthawk *Podager nacunda minor*                                       30cm

Large and bulky with narrow wings and shortish tail. Leisurely flight, often glides with wings held in dihedral. Adult male has brown head, mottled black and white. Throat white contrasting with brown upper-breast band. Rest of underparts white. Upperparts brown, mottled black and white, with outer tail feathers tipped white. Flight feathers black with broad central white band. In flight, underwing-coverts conspicuously white. Female lacks white in outer tail. Juvenile recalls adult. **Voice** Rarely heard. **SS** Unlikely to be confused. **Status** Rare visitor to Trinidad with most sightings Jun–Oct. Flies high over lowland savannas, cultivated land and scrub, often in small groups. Has bred. No records on Tobago for many years.

Pauraque

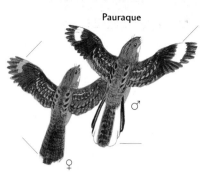

♂

♀

Lesser Nighthawk

♂

Short-tailed Nighthawk

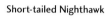

Nacunda Nighthawk

# SWIFTS – APODIDAE

Swifts are almost entirely aerial, landing only to nest. They are difficult to identify due to their rapid flight and similar plumages; wing shape and general proportions are key factors.

## Chestnut-collared Swift *Streptoprocne rutila rutila* 14cm

Medium-sized, all dark, with long, tapered wings and fairly long, square-ended tail. Adult male is entirely dark sooty-brown with broad, complete, reddish-brown collar that is surprisingly difficult to see against a bright sky. Female has much-reduced or no collar; entirely lacking in juvenile. **Voice** Guttural, high-pitched chatter. **SS** From *Chaetura* swifts by larger size, longer more powerful wings and all-dark rump and throat. From larger White-collared by lack of white collar. **Status** Locally common Trinidad resident, mostly in Northern Range, but frequently seen at lower altitudes, especially in wet season. Absent from Tobago.

## White-collared Swift *Streptoprocne zonaris albicincta* 20cm

Large and rakish with sickle-shaped wings and notched tail. Adult entirely black with complete white neck collar, that is narrower or broken on female. Immature more sooty-black with vestigial collar. **Voice** Squeaks, whistles and buzzes, including faint but harsh *zee zee zee*. **SS** None. **Status** Seasonal visitor to Trinidad, common Jul–Oct, when early-morning and late-afternoon flocks often number hundreds over Northern Range. Occasional Feb–Jun but no recent winter records. Three Aug–Oct records for Tobago.

## Band-rumped Swift *Chaetura spinicaudus spinicaudus* 11cm

Slender, narrow-winged and rather long-tailed. Throat pale grey, below black. Upperparts black with square white rump. **Voice** Thin and sweet, high-pitched *tsip tsip tsip*. **SS** White rump eliminates other small swifts. **Status** Common over forest, particularly edges, usually above 150m on Trinidad. Absent Tobago.

## Grey-rumped Swift *Chaetura cinereiventris lawrencei* 11cm

Slender; sickle-shaped wings and long tail. Below mid-grey; undertail-coverts black. Black above with rump and uppertail-coverts mid-grey, tail black. **Voice** High-pitched rather maniacal chattering. **SS** From larger Chapman's by black undertail-coverts/undertail; from Band-rumped by rump; and Short-tailed by wing shape. **Status** Common Trinidad resident over hill forest, usually above 300m. Less common Tobago.

## Short-tailed Swift *Chaetura brachyura brachyura* 10cm

Flight rather bat-like and slow compared to other *Chaetura*. Pale buff-grey lower back, rump and tail. Race *praevelox* on Tobago browner with paler throat. **Voice** Frenetic twittering and cricket-like *stut stut stut*. **SS** Easiest of genus to identify as wings appear broad-based and blunt, affording paddle-shaped profile. Leading edge to wing curved but trailing edge straight. Tail exceedingly short. From Grey-rumped by darker throat, buff 'rear half', wing and tail shape. From Chapman's by shape, darker underparts and strongly contrasting upperparts. **Status** Very common resident on T&T, especially over lowland and suburban areas.

## Chapman's Swift *Chaetura chapmani chapmani* 13cm

Appears uniformly dark. Blackish head barely contrasting with dark grey throat. Underparts dark grey-brown. Nape and wings blackish, mantle, rump and tail dark grey-brown. **Voice** Drawn-out *che'e'e'ed*, with various ticking notes. **SS** Smaller than female Chestnut-collared Swift, with narrower wings and tail. From Short-tailed by wing shape, rump and tail colour. **Status** Probably a scarce resident on Trinidad; feeds over forests of Northern Range and central savannas. Absent from Tobago.

## Lesser Swallow-tailed Swift *Panyptila cayennensis cayennensis* 13cm

Narrow, pointed wings and long, forked tail. All black with conspicuous white collar and prominent white rump-sides. Tail fork usually only visible when banking. **Voice** Usually silent. **SS** None. **Status** Widespread, yet uncommon resident on Trinidad. Absent Tobago.

## Fork-tailed Palm Swift *Tachornis squamata squamata* 13cm

Slender body, very narrow wings and long, forked tail. Dark brown crown. Face, throat to undertail-coverts white, flanks dark brown. Upperparts dark brown; flight feathers darker. Tail thin and tapered, forked tips often concealed. **Voice** A quiet buzz. **SS** From Lesser Swallow-tailed by even-slimmer body, brown upperparts and extensive white underparts. **Status** Locally common Trinidad resident, favours open lowlands and suburban areas, where it is a low-level feeder. Just one recent record for Tobago.

Chestnut-collared Swift

juvenile

♂ adult

White-collared Swift

juvenile

adult

Band-rumped Swift

Grey-rumped Swift

Lesser
Swallow-tailed Swift

Short-tailed Swift

Chapman's Swift

Fork-tailed Palm Swift

## HUMMINGBIRDS – TROCHILIDAE

Seventeen species recorded in T&T. Typically have dazzling metallic plumage, with a great variety of bill shapes and tail-length. Some older females acquire partial male plumage. Often tame and confiding. Most are resident, but there is an obvious post-breeding dispersal to mainland South America in some species.

### Rufous-breasted Hermit *Glaucis hirsutus insularum* 13 cm

Large with long, decurved bill and rounded tail. Dark olive crown, blackish face with indistinct buff supercilium and malar. Chin to breast dirty rusty, grey on belly. Olive-green above with uppertail-coverts fringed buff. Tail rounded, rufous with black subterminal band and white tip. Upper mandible black, lower yellow. Juvenile recalls dull adult. **Voice** Thin rather drawn-out *tssip* or disyllabic *swee-it* in flight; song rarely heard. **SS** From Green and Little Hermits by rounded tail without long central feathers, size and underparts colour. **Status** Common resident in forest on both islands, especially around *Heliconia* plants.

### Green Hermit *Phaethornis guy guy* 17 cm

Large and slender with very long, decurved bill and long central tail feathers. Adult male has dark grey-green head with broad black comma-shaped mask, thin buff supercilium and buff moustachial. Underparts grey-green with thin buff throat-stripe and white-fringed undertail-coverts. Upperparts dark green with blue tone to rump. Flight feathers black. Tail dark green with white central feather tips. Upper mandible black, lower mostly red. Female has more curved bill and clear buffy postocular, malar and throat stripes; young has buffy fringes to upperparts. **Voice** Contact a low-pitched *squa-ik*, repeated every few seconds. Male's lek call is a repetitive nasal *hu-ek hu-ek hu-ek*. **SS** From Rufous-breasted by long central rectrices and underparts colour. From smaller Little by overall colour. **Status** Fairly common Trinidad resident, in cultivations, second growth and forest at all elevations. Absent from Tobago.

### Little Hermit *Phaethornis longuemareus longuemareus* 10 cm

Small with long, slightly decurved bill and long white central rectrices. Adult has bronze-brown head with buff supercilium broadening behind eye, black mask, buff moustachial and thin black loral stripe. Variable blackish bib; upper breast orange-buff, paler on belly. Upperparts buff-brown faintly streaked black, becoming rufous on rump. Wings sooty-black. Upper mandible black, lower yellow with dark tip. Female smaller with less well-developed bib. Young have rufous fringes to upperparts. **Voice** Sweet *swee-ee*, repeated severally, and a high-pitched squeak. **SS** See Green Hermit. **Status** Locally common Trinidad resident in lower- and mid-elevation forests, especially at edges. Absent from Tobago.

### White-tailed Sabrewing *Campylopterus ensipennis* 13 cm

Large and long-winged with long, slightly decurved bill and rounded tail. Adult male has green crown, blackish face and small white postocular spot. Throat dark blue, rest of underparts green. Undertail appears all white at rest. Upperparts green with black wings. Outer three rectirces white, central tail darker green. Bill all black. Female duller with blue-green throat, thin white malar and grey central underparts spangled green. Immature like adult female. **Voice** High-pitched, fast chatter, *chit-it chit-it* at lek, and single high-pitched *chirp*, and *ker-swit*. **SS** None. **Status** Uncommon resident confined to forested Main Ridge on Tobago.

### Brown Violetear *Colibri delphinae* 12 cm

Fairly large and chunky, with short square tail and short straight bill. Grey-brown head with dark purple ear-coverts. Throat iridescent blue-green, with white sides. Below underparts grey-brown with orange-buff undertail-coverts. Darker above, buff rump/uppertail-coverts, and darker terminal tail-band. Bill black. Young have head brown, but may show emergent blue platelets. **Voice** Loud strident *tsik tsik tsik*. **SS** None. **Status** Scarce resident, favouring forested Northern Range and NE Trinidad. Just one breeding record for Tobago.

### Ruby Topaz *Chrysolampis mosquitus* 9 cm

Small but distinctly large-headed, often with shaggy-crested look. Short, fairly square tail, usually closed at perch. More than most hummingbirds, male looks very dull to blackish in some light lights, but extremely colourful in others. Bright orange front and cerise hindcrown make adult male unmistakable if seen well, but note range of plumages. Female green above, dull creamy below with irregular line of platelets on throat, chestnut base to tail, broad dark subterminal band and white tips. Juveniles sexed by tail pattern, whilst male often has some orange on throat. **Voice** Short, sharp buzzing *tzik*. **SS** None. **Status** Fairly common breeding visitor to both islands, with most departing to mainland Sep–Dec.

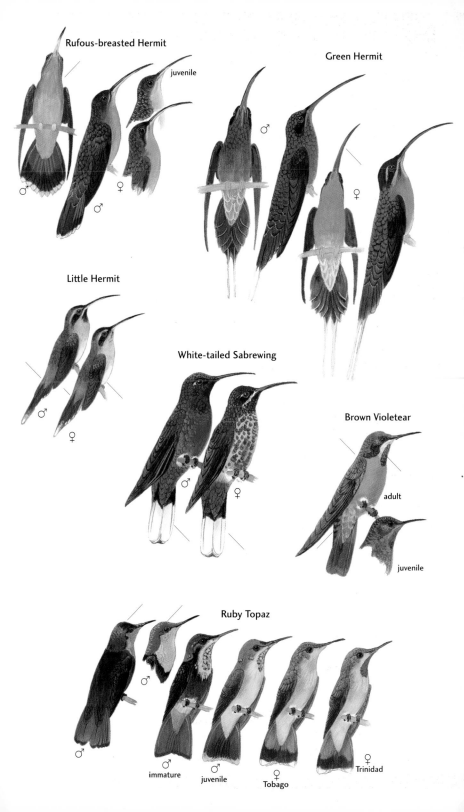

**Rufous-breasted Hermit**

juvenile

♂

♂

♀

**Green Hermit**

♂

♀

**Little Hermit**

♂

♀

**White-tailed Sabrewing**

♂

♀

**Brown Violetear**

adult

juvenile

**Ruby Topaz**

♂

♂

immature

♂

juvenile

♀

Tobago

♀

Trinidad

## Green-throated Mango *Anthracothorax viridigula*                                  13cm

Large with long, slightly decurved bill and square tail. Adult male has emerald-green head, throat and breast-sides. Central breast and belly black. Undertail purple with black outermost feather. Back bottle green, lower mantle and rump variably bronzy-green. Central tail feathers black, rest purple. Wings black. Bill black. Female has copper-green head with darker mask and obscure white postocular spot. Central underparts white with black central stripe and green flanks. Upperparts bronzy-green. Central tail black, rest wine-red, narrowly tipped white. Juvenile recalls adult female with buffy flanks. **Voice** Fast series of hard, tapping notes. **SS** Male separated in good light from Black-throated Mango by throat colour and lack of blue breast-sides. Female with extreme care by variable shorter black breast-stripe, reddish crown and upperparts sheen, and white outertail tips. **Status** Uncommon Trinidad resident in mangroves and cultivation with blooming *Immortelle*. Absent from Tobago.

## Black-throated Mango *Anthracothorax nigricollis nigricollis*                     11cm

Medium-sized with long, slightly decurved bill and square tail. Adult male has apple green head, darker around eyes. Throat and breast black, bordered by blue stripe. Black extends as central belly-stripe with green flanks. Back apple green becoming golden-green on lower mantle and rump. Central tail feathers black, rest deep wine-red, narrowly tipped black. Wings black. Bill black. Female has bottle green head with blackish mask and white postocular spot. Underparts white with broad yet ragged black central stripe. Upperparts glossy bottle green, flight feathers black. Juvenile recalls female but has broken, narrower black stripe on underparts and some white feathers in green mantle. **Voice** A sharp *chick*. **SS** Male separable in good light from Green-throated Mango by throat and breast colour. Female with care by usually broader, longer black stripe on underparts. **Status** Common Trinidad resident at all elevations, less numerous on Tobago. Favours suburban gardens, mangrove edges, secondary scrub and forest edge. Many depart for mainland in Sep–Dec.

## White-necked Jacobin *Florisuga mellivora*                                        11cm

Medium-sized with almost straight bill and rounded tail. Adult male has indigo-blue head, darker around eyes. Throat and upper breast bright blue-green, rest of underparts white. Nuchal collar white, rest of upperparts and central tail feathers bottle green. Outer tail feathers white, narrowly tipped black and often spread when hovering. Wings black. Bill black. Female has forecrown, face, throat and breast-sides blue-green, fringed white with heavy black scales. Belly dirty white, undertail-coverts dark grey. Rear crown and upperparts green. Tail dark blue, tipped white. Flight feathers black. **Voice** Weak, high-pitched chatter, *dzit dzit dzit*. **SS** Unmistakable. **Status** Locally common resident in forest and cultivated estates, at all elevations on Trinidad. Some may move to mainland in Oct–Dec. On Tobago, found mainly in Main Ridge forests and woodland in SW.

## Blue-chinned Sapphire *Chlorestes notata notata*                                  9cm

Small and straight-billed with rounded tail. Adult male has bright green head. Throat bright blue; rest of underparts bright green, with white thighs. Upperparts green. Wings and tail deep royal blue. Bill straight with black upper and reddish lower mandibles. Female has duller green head. Breast dirty white with patchy green blotches. Undertail-coverts green. Upperparts green with dark blue wings and tail. **Voice** Extremely high-pitched hiss. **SS** Adult male from Copper-rumped Hummingbird by green rump. From slightly smaller Blue-tailed Emerald by reddish lower mandible and blue chin. Female from White-chested Emerald by less extensive white underparts, bill colour, green rump and blue tail. **Status** Common and widespread Trinidad resident, prefers forest and forest edge. Very rare on Tobago.

## Blue-tailed Emerald *Chlorostilbon mellisugus caribaeus*                          7.5cm

Tiny. Short, straight bill and slightly forked tail. Adult male has head and underparts bright green, with white thighs. Upperparts slightly darker green. Wings black, tail steel blue. Bill short and black. Female has green head with black ear-coverts and white postocular supercilium. Underparts grey-white. Upperparts green, wings black. Tail steel blue with outer feathers tipped white. Juvenile like adult female but has buff underparts. **Voice** Thin, high-pitched *seep*. **SS** Male from Blue-chinned Sapphire by black bill; green throat only distinctive in sunlight. Female unlikely to be confused. **Status** Locally common resident on Bocas Is and NW peninsula of Trinidad, rare further east and absent from Tobago.

## Green-throated Mango

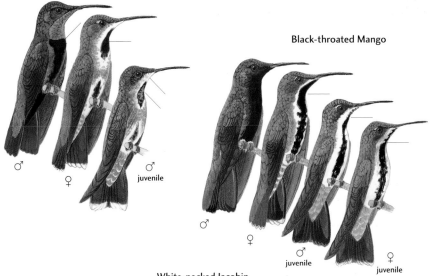

♂

♀

juvenile

## Black-throated Mango

♂

♀

juvenile

juvenile ♀

## White-necked Jacobin

♀
juvenile

♀

♂
juvenile

♂
immature

adult ♂

## Blue-chinned Sapphire

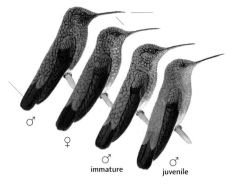

♂

♀

immature ♂

juvenile ♂

## Blue-tailed Emerald

♂

♀

juvenile

## White-tailed Goldenthroat *Polytmus guainumbi guainumbi*                    10cm

Medium-sized, with long, slightly decurved bill and long, rounded tail. Adult male has dark brownish-green crown. Face grey with black hermit-like tear mark from lores to ear-coverts, bordered either side by white line. Throat to belly green, rest scalloped green and white. Nape and mantle frosted green, with rufous tone to rump. Central tail green, outer tail feathers broadly tipped and fringed white. Culmen and tip black, rest of bill red. Female spangled green on chin to breast, creamy belly and flanks, rest white. Young male green from chin to central belly, breast-sides creamy, rest white. Young female creamy from chin to belly, rest white. **Voice** Series of high-pitched *sip* notes and squeaks. **SS** None. **Status** Uncommon Trinidad resident, though much scarcer in second half of year. Found in lowland grassland, scrub and marsh edge. Absent from Tobago.

## White-chested Emerald *Amazilia chionopectus chionopectus*                    9.5cm

Fairly small, almost straight-billed and slightly fork-tailed, with small postocular spot. Entirely metallic green above, usually with golden reflections but varies and may even be purpurescent. All white below, with scattered green platelets on sides and flanks. Tail golden-brown. **Voice** High-pitched series of churrs, squeaks and buzzes. **SS** From female Blue-chinned Sapphire by all-black bill and golden tone to rump and tail. From female Blue-tailed Emerald by lack of supercilium, ear-patch and white in outer tail, and golden rump. **Status** Common and widespread Trinidad resident. Absent from Tobago.

## Copper-rumped Hummingbird *Amazilia tobaci erythronotus*                    8cm

Small, straight-billed with slightly forked tail. Adult has coppery-green crown. Face and underparts bright green with white rear flanks, thighs and undertail-coverts rufous. Nape and fore mantle coppery-green, becoming golden on shoulder; lower mantle and rump coppery-bronze, uppertail-coverts purpurescent. Wings and tail dark royal blue. Bill short and mainly black with reddish base to lower mandible. *A. t. tobaci* (Tobago) is larger with greener back and reddish-chestnut uppertail-coverts. **Voice** From song-perch, *tee, tu tu tu*, fading towards finish. Also, an incisive *chit*. **SS** From male Blue-chinned Sapphire by chin colour and copper-bronze lower mantle and rump. From Blue-tailed Emerald by mantle and rump colour. **Status** Very common resident that is widespread on both islands.

## Long-billed Starthroat *Heliomaster longirostris longirostris*                    11cm

Medium-sized with very long, straight black bill and slightly forked tail. Adult male has azure-blue crown. Face olive-green with white malar. Throat raspberry red. Breast greyish-olive, belly and leg-puffs grey-white. Upperparts, including central tail, bronze-green with ill-defined white line on rump. Outer tail black tipped white. Wings black. Female has green crown, white postocular spot, darker throat and white tail tip. Juvenile has buff malar. **Voice** Quiet *swip swip*. **SS** None. **Status** Scarce Trinidad resident. Present year-round, but harder to find Sep–Nov. Absent from Tobago.

## Tufted Coquette *Lophornis ornatus*                    7cm

Tiny with short, straight bill, and notched tail. Long, shaggy crest; long, orange cheek plumes with metallic green spots, throat appears dark. Underparts green, with narrow orange breast-band. Upperparts green with golden reflections from lower back to outer tail, pale band on rump. Bill vermilion, tipped black. Female lacks crest and plumes, with orange-buff face, throat and upper breast. Juvenile like duller female, with white-speckled throat. **Voice** Rarely heard; occasionally a faint *tzik*. **SS** Male unmistakable. Female and immature from Rufous-shafted Woodstar by rump band and lack of white postocular spot. **Status** Fairly common Trinidad resident, favours forest edge and Vervain hedges. Absent from Tobago.

## Rufous-shafted Woodstar *Chaetocercus jourdanii jourdanii*                    7cm

Tiny with short, straight bill; male has uniquely shaped tail. Breeding male has green head with short white postocular line. Throat purple bordered by white pectoral band that reaches neck-sides. Rest of underparts dull green with white flank mark, most noticeable when hovering. Upperparts green. Bill black. In eclipse, throat and diffuse breast-band buff. Female has dull green head with blackish ear-coverts and white postocular spot, rest of underparts cinnamon-buff. Upperparts green, tail cinnamon with black bar. Juvenile like female, but male washed brownish on rump to tail. **Voice** Thin, *tssit, tssit*. **SS** Male unmistakable. See Tufted Coquette. **Status** Very rare visitor to Trinidad from mainland, with most records Mar–Aug. Favours flowering shrubs adjacent to hill forest edge. Absent from Tobago.

## White-tailed Goldenthroat

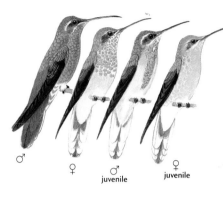

♂            ♀
      juvenile        juvenile
         ♂           ♀

## White-chested Emerald

extremes of variations

## Copper-rumped Hummingbird

*tobaci*          *erythronotus*
(Tobago)          (Trinidad)

## Long-billed Starthroat

♂          ♀

## Tufted Coquette

♂

♀

juvenile

## Rufous-shafted Woodstar

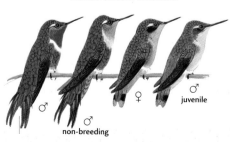

♂          ♀          ♂
      non-breeding        juvenile
           ♂

# PLATE 67: TROGONS

## TROGONS – TROGONIDAE

Forty species that range through Asia, Africa and the New World. Only three occur on Trinidad, two also on Tobago. Characterised by large head, short wings and long tail, these brightly coloured forest birds often sit motionlessly in mid canopy. Vocalisations similar amongst these three but vary in pitch, number of notes and speed of delivery.

### Amazonian Violaceous Trogon *Trogon violaceus violaceus*                    22cm

Adult male has blue-black head and thin yellow orbital ring. Upper breast dark blue-green, rest of underparts bright yellow. Undertail white barred black. Upperparts deep bottle green washed blue on tail. Wings black, finely vermiculated white on coverts. Bill greenish-white, legs grey. Female has grey head with incomplete whitish orbital restricted to sides. Throat and upper breast grey, rest of underparts yellow. Upperparts and tail grey, with small white coverts spotting. Flight feathers blackish. Bill grey. **Voice** Rapid series of 20+ *kiu* notes, accelerating at end. Also, a guttural *kraa kraa kraa*. **SS** Male from Amazonian White-tailed by smaller size, yellow orbital and barred undertail. Female by smaller size and incomplete orbital. **Status** Common Trinidad resident, overlapping in habitat with Amazonian White-tailed Trogon. Several undocumented reports from forest on Main Ridge, Tobago.

### Collared Trogon *Trogon collaris exoptatus*                    25cm

Adult male has dark green crown and blackish face with thin yellow or reddish orbital. Throat and upper breast bottle green bordered by narrow white pectoral band. Rest of underparts red. Undertail heavily barred black and white with three broader white bands. Upperparts bottle green, brighter on rump with black terminal tail-band. Wing-coverts and shoulder vermiculated black and white; primaries black with white shafts. Bill pale yellow, legs grey. Female has brown head, darker around face and broad white eye brackets. Throat and upper breast brown with thin white pectoral band. Rest of underparts pinkish-red. Undertail white, densely barred black. Mantle brown, more rufous on rump. Tail brown with white outer feathers. Wing-coverts grey, flight feathers black. Bill grey. Immature brown and buff, with dull red restricted to undertail-coverts. **Voice** Usually 3–4 notes, rarely 5–6: a disyllabic *kiow kiow kiow*, slightly higher pitched than Amazonian White-tailed but with same speed of delivery. Also, a low-pitched rolling *churrr*. **SS** Unmistakable. **Status** Uncommon resident in T&T, favouring forest usually above 200m.

### Amazonian White-tailed Trogon *Trogon viridis viridis*                    28cm

Adult male has dark blue-black head with bold blue-white orbital ring. Throat and upper breast dark blue; rest of underparts yellow. Undertail unmarked white. Upperparts dark blue-green with narrow black terminal tail-band. Wings short and black. Upper mandible silver-grey, lower yellow-grey. Legs grey. Female has sooty-black head and blue-white orbital. Chin, throat and upper breast grey; lower breast bright yellow, becoming orange-yellow on lower belly/undertail-coverts. Undertail white with narrow black barring and solid white tips. Upperparts dark grey. Wings dusky with faint black and white vermiculations; white primary shafts conspicuous. Bill blackish. **Voice** Series of 6–18 notes, *kiou kiou kiou...*, slower in delivery than Amazonian Violaceous, longer and slightly lower pitched than Collared. Also, a quick hollow *chup chup chup*. **SS** Adult male from Amazonian Violaceous by larger size, blue-white orbital and white unmarked undertail. Female by larger size and complete orbital. **Status** Common Trinidad resident, favouring forest at all elevations. Absent from Tobago.

## Amazonian Violaceous Trogon

♂    ♂
     immature    ♀

## Collared Trogon

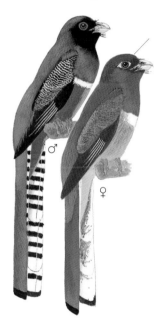

♂

♀

## Amazonian White-tailed Trogon

♂

♀

♂
immature

## MOTMOTS – MOMOTIDAE

Motmots are usually seen perched quietly on a horizontal branch in the subcanopy, or surprisingly near the ground, with the tail swinging from side to side in a slow, deliberate manner, but accelerating the movements when alert. The tail feathers lack the rackets when they first emerge, but there is a section near the end where the barbs are quite weak and they wear off.

### Blue-crowned Motmot *Momotus momota bahamensis*     46cm

Adult has black crown bordered by broad azure-blue stripe from bill to rear crown, broad black lores and mask, reddish-brown eyes and blue ear-coverts. Underparts rufous-brown with some isolated blue feathers on upper breast. Upperparts grey-green with buff wash to nape. Tail blue-green. Bill stout and black. Legs black. Juvenile like adult, but lacks blue breast feathers. **Voice** Soft but far-carrying *whoop*, occasionally two or more notes. **SS** Unmistakable. **Status** Shy and uncommon Trinidad resident of mid- to high-elevation forests. More common and widespread on Tobago.

## JACAMARS – GALBULIDAE

Predominantly shiny, metallic green, with long, scissor-like bills and long tails, jacamars feed on insects in flight and are particularly adept at catching butterflies.

### Rufous-tailed Jacamar *Galbula ruficauda ruficauda*     27cm

Adult male has a bottle green head with black lores and eye-patch. Chin and throat white. Upper-breast band bottle green, appearing golden in bright sunlight; rest of underparts brick red. Upperparts green with golden sheen to wing-coverts; flight feathers black. Central tail feathers dark green, rest rufous-brown. Bill black, feet yellow. Female has buff throat. Immature recalls adult but is duller. **Voice** A shrill, piercing *preee*, often quickly repeated with last two notes slightly higher pitched. **SS** Unmistakable. **Status** Uncommon Trinidad resident of forest edge and second growth at all elevations. Much more common on Tobago.

## TOUCANS – RAMPHASTIDAE

Toucans are birds of the Neotropical rainforests, and only one species occurs in our region, on Trinidad. They are omnivorous, but are particularly fond of, and adept at, plucking stone-fruits, like date palms.

### Channel-billed Toucan *Ramphastos vitellinus vitellinus*     51cm

Adult has black crown and white face, with pale blue in front of and around eyes. Neck-sides and upper breast white, central breast orange, lower breast deep red. Belly black, undertail-coverts red, undertail black. Upperparts black with red uppertail-coverts. Bill black with pale blue base. Legs blue-grey. Characteristic undulating flight pattern, with bill weight forcing head and neck below the horizontal. Juvenile like dull adult but with smaller bill. **Voice** Single, repeated far-carrying *ki-aarh*. **SS** Unmistakable. **Status** Common Trinidad resident in secondary scrub, cultivated estates and forest. Absent from Tobago.

**Blue-crowned Motmot**

**Rufous-tailed Jacamar**

♂

♀

**Channel-billed Toucan**

# KINGFISHERS – ALCEDINIDAE

Whilst 93 species occur worldwide, only six are found in the New World. On Trinidad, five species have been recorded, of which three are resident, one a winter visitor from North America and the last a vagrant from the south. On Tobago, two species occur. All have disproportionately large heads, dagger-like bills, and short necks and wings.

## Ringed Kingfisher *Megaceryle torquata torquata* 41 cm

Huge and shaggy-crested. Adult male has blue-grey head and double crest, white loral spot and broad, complete white neck collar. Underparts to belly brick red with white undertail-coverts. Upperparts blue-grey with darker flight and outer tail feathers, spotted white. Underwing-coverts white. Bill basally yellow, distally black. Legs yellow-grey. Female has broad blue-grey and narrow white breast-bands, brick red underwing linings and undertail-coverts. Juvenile darker above and heavily spotted, with dark streaks on back; breast is mottled and belly paler red. **Voice** A loud single sharp *kek*; in alarm, repeated as 7–8-note rattle. **SS** From Belted Kingfisher by much larger size, two-toned bill, rufous underparts, female by reddish underwing-coverts. **Status** Common resident in C & W Trinidad, favouring mangroves and flooded agricultural land. One recent record on Tobago, Jul 2007.

## Belted Kingfisher *Megaceryle alcyon* 33 cm

Large and shaggy-crested. Adult male has blue-grey head and double crest, white loral spot and broad, complete white neck collar. Underparts white with broad blue-grey breast-band. Upperparts blue-grey with white wing-coverts spots. Flight feathers dark grey with large white patch at base of outer primaries. Underwing-coverts white. Bill stout and all dark. Legs black. Female has chestnut lower breast-band that extends onto flanks. Immature recalls adult, but male has tawny breast-band heavily flecked black; female additionally has narrow and broken rufous lower breast-band. **Voice** A loud *kek kek kek*. **SS** Much like a small version of Ringed Kingfisher, but separable by wholly dark bill and extent of white on underparts at all ages. **Status** Uncommon winter visitor to T&T; present early Oct–mid-Apr in coastal areas of N and E Trinidad and more widespread on Tobago.

Ringed Kingfisher

♂

♀

juvenile

Belted Kingfisher

♂
immature

♂

♀

♀
immature

# PLATE 70: SMALLER, GREEN KINGFISHERS

### Amazon Kingfisher *Chloroceryle amazona* 28cm

Large with a moderately shaggy crest. Adult male has bottle green crown and rear ear-coverts; frontal face black. White throat extends to form complete collar. Breast and belly white with broad, brick red upper-breast band and bottle green flanks streaking. Undertail-coverts white, undertail black with row of white spots down centre. Upperparts dark bottle green. Bill large and black; legs dark grey. Female has incomplete bottle green breast-band. Immature recalls adult female, with extensive buff spotting on wing-coverts and yellow on lower mandible; male shows some red on sides of breast. **Voice** A loud rattling *kikikikiki*, with harsh descending trill. **SS** Adult from Green Kingfisher by much larger size and unspotted wings. **Status** Accidental on Trinidad, with just two old records, one a breeding record from forested Madamas River.

### Green Kingfisher *Chloroceryle americana croteta* 19cm

Small and plump. Adult male has dark bottle green head with darker lores. Throat white continuing as nuchal collar, with very thin black moustachial. Breast and belly white with broad chestnut upper-breast band and dark green spots on lower breast and flanks. Upperparts dark bottle green with white wing-covert tips and outer tail feathers. Flight feathers black. Bill long and black; legs black. Female has buff throat and upper breast, and two spotty olive-green breast-bands. Juvenile recalls respective adult but is duller with extensive buff crown and spotting on wing-coverts. **Voice** A loud, sharp and grating *tchik* and hard rattle. **SS** From American Pygmy Kingfisher by larger size and mainly white underparts. From Amazon Kingfisher by much smaller size, lack of a shaggy crest, white wing-coverts spotting, and spotted flanks. **Status** Common resident in T&T, found mainly in lowland freshwater marshes and brackish mangroves. Occasionally seen near hill forest streams

### American Pygmy Kingfisher *Chloroceryle aenea* 14cm

Tiny and short-winged. Adult male has bottle green head with orange loral spot. Throat dull orange continuing as nuchal collar. Breast and flanks orange, central belly to undertail-coverts white. Upperparts bottle green. Bill black, legs brownish-grey. Female has dark green breast-band. Juvenile recalls adult female, with duller underparts and conspicuous buff spotting on wing-coverts. **Voice** A dry and high-pitched buzz. **SS** From Green Kingfisher by much smaller size, extensive orange underparts and lack of white wing spots (adult). **Status** Uncommon Trinidad resident, usually in brackish mangrove swamps, but very occasionally near forested streams of Northern Range, to 400m. Perches motionless for long periods, usually no more than 1m above water. Absent from Tobago.

Amazon Kingfisher

♂

♂
immature

♀

Green Kingfisher

American Pygmy Kingfisher

♂

♀

♀

juvenile

♂

# WOODPECKERS – PICIDAE

Five woodpeckers occur on Trinidad, of which two also occur on Tobago with a sixth only found on the latter. Variable in size and coloration, but all share the characteristic undulating flight action.

### Red-crowned Woodpecker *Melanerpes rubricapillus rubricapillus*     18cm

Forecrown buff, crown red, face unmarked olive-grey. Underparts slightly darker with reddish stain to lower belly. Nape dull red. Mantle and wings densely barred black and white. Rump and uppertail-coverts form prominent square white patch. Central tail feathers white barred black; outer tail black. Bill and legs dark grey. Female lacks red crown. Immature lacks red crown and reddish belly, but has red stain on nape and brown and buff mantle barring. **Voice** A raucous, raspy rattle *krrrrrr*. **SS** From other woodpeckers by white rump, barred mantle and wings. **Status** Common, widespread Tobago resident, particularly fond of coconut estate, but not normally above 300m. Absent from Trinidad.

### Red-rumped Woodpecker *Veniliornis kirkii kirkii*     17cm

Adult male has red crown with bland olive-grey face. Underparts grey-white, densely barred olive. Nape, mantle and wings olive-green with thin yellowish collar and faint buff wing-coverts spotting. Rump red; tail black. Bill and legs black. Female has browner crown and nape. Immature recalls duller adult. **Voice** Series of *blik blik blik* notes repeated 5–6 times. **SS** From Golden-olive Woodpecker by slightly smaller size, bland face, lack of broad moustachial and mantle and rump colour. **Status** Uncommon resident in T&T, found in mangrove and forest on Trinidad, and Main Ridge forest on Tobago.

### Golden-olive Woodpecker *Piculus rubiginosus trinitatis*     19cm

Adult male has slate-grey forecrown; rear crown red. Face yellowish-white with black eyes, affording startled expression, dark grey barring on rear ear-coverts/neck-sides and broad red moustachial stripe thinly bordered black. Chin and throat dark grey. Underparts pale yellow barred dark olive, most intensely on upper breast. Nape red. Mantle and rump golden-brown with darker olive-brown wings and tail. Bill and legs dark grey. Female lacks red on crown and face. Immature recalls duller adult. Tobago race *tobagensis* is larger, has heavier bill and broader, more greenish bars on breast. **Voice** A single harsh, nasal *hu-eek*, sometimes followed by low-pitched rattling trill. **SS** See Red-rumped Woodpecker. **Status** Common resident in T&T, in wooded estates and forest at all elevations.

### Chestnut Woodpecker *Celeus elegans leotaudi*     28cm

Triangular crown is creamy yellow and extends in pointed crest. Face cinnamon-brown with diffuse, broad greyish orbital and short, broad red malar. Chin to upper belly cinnamon-brown, rest creamy yellow. Back cinnamon-brown, wing-coverts sparsely dotted white. Flight feathers darker brown. Rump creamy yellow, tail black. Bill ivory, legs grey. Female lacks red malar. Immature has darker face. **Voice** A grating, harsh *kee-aa*. **SS** None. **Status** Locally uncommon Trinidad resident in Northern Range; scarce elsewhere and absent from Tobago.

### Lineated Woodpecker *Dryocopus lineatus lineatus*     33cm

Adult male has red triangular-shaped head to eye level, black facial skin on cheeks and ear-coverts, a thin buff-white moustachial running from base of bill to neck-sides, and a broad red malar wedge. Chin buff densely streaked black. Throat and upper breast black, lower breast and belly pale yellow becoming buff-white on flanks, heavily barred black. Upperparts black with inner scapulars white, forming two almost parallel white lines on mantle-sides. Underwing-coverts white. Bill yellowish-grey, legs grey. Female lacks red malar and has black forecrown. Immature duller, often with more black on breast. **Voice** Explosive series of notes on same pitch, *wuk wuk wuk wuk*, and a single sharp *pik*. **SS** See Crimson-crested. **Status** Common, widespread Trinidad resident; absent from Tobago.

### Crimson-crested Woodpecker *Campephilus melanoleucos melanoleucos*     36cm

Male has red head, pale yellow lores, conspicuous white lower-ear-coverts spot and broad white stripe on neck-sides. Throat and upper breast black; rest of underparts pale yellow densely barred black. Upperparts sooty-black, with broad white stripe over scapulars forming V on mantle. Underwing-coverts white. Bill yellowish-grey, legs grey. Female has black central forehead and broad white moustachial reaching neck-sides. Immature duller and browner. **Voice** A short, rolling *aarrh*. **SS** From Lineated by extent of red on face and by scapular lines meeting on mantle. **Status** Uncommon resident Trinidad lowlands. Absent from Tobago.

Red-crowned Woodpecker

♀

♂

Red-rumped Woodpecker

♀

♂

Golden-olive Woodpecker

♂

♀

Lineated Woodpecker

♀

♂

Crimson-crested Woodpecker

♀

♂

Chestnut Woodpecker

♀

♂

# OVENBIRDS – SUBFAMILY FURNARIINAE

Some 240 species of this diverse family occur throughout the Neotropics. Spinetails have long slender bodies, short wings, thin and long ragged tails and often very secretive habits. There are 63 species, with just three resident on Trinidad, each favouring its own distinct habitat. One also occurs on Tobago. Xenops are characterised by their small size, slightly upturned bill, rufous wing-stripe and nuthatch-like feeding actions. They are represented by five species with just one on Trinidad. Leaftossers are found almost exclusively on the forest floor and characterised by their strong bills, short wings, short legs and pot-bellied appearance. Of the six species, just one occurs in T&T.

### Pale-breasted Spinetail *Synallaxis albescens trinitatis* 17cm

Adult has chestnut crown. Face olive-grey with indistinct black eye-stripe. Throat off-white with thin black stripe. Breast and belly dirty olive-grey, undertail-coverts and rear flanks warmer. Upperparts olive-grey with chestnut wing-coverts and brownish-olive flight feathers and tail. Bill and legs grey. Young birds dull olive-brown. **Voice** Rasping *wee chirr* with emphasis on second note. Also, a rather metallic *chack*, usually from dense scrub. **SS** From Stripe-breasted Spinetail by overall grey plumage, chestnut cap and wing-coverts, and unstreaked throat. Pale-breasted prefers dry scrub. **Status** Uncommon skulking resident of lowlands, open scrub and savanna on Trinidad. Absent on Tobago.

### Stripe-breasted Spinetail *Synallaxis cinnamomea carri* 15cm

Adult has dark reddish-brown crown, greyer cheeks and ear-coverts. Chin white, throat black densely speckled and streaked white. Rest of underparts olive-brown. Mantle and rump dark brown, with rufous-chestnut wings. Tail olive-brown. Bill grey with dusky tip, legs buffy grey. Immature brighter than adult and somewhat scalloped below. *S. c. terrestris* (Tobago) has paler face and throat, buff breast streaked black, and pale grey belly and undertail-coverts. Juvenile brighter below and scalloped on breast, sides and flanks. **Voice** Repetitive two-syllable *si boyee*, a *tcheu tcheu* and quick *chee-chee-chee* in alarm. **SS** See Pale-breasted Spinetail but Stripe-breasted prefers forest and coffee plantations. **Status** Common resident in T&T. Keeps low to forest floor, foraging in tangles and leaf litter, and best located by call.

### Grey-throated Leaftosser *Sclerurus albigularis zamorae* 17cm

Adult has dark brown crown contrasting with grey face and slightly paler throat. Breast warm rufous-orange, becoming dark brown on belly and undertail-coverts. Nape, mantle and wings dark brown. Extensive dark red rump contrasts with black tail. Bill dark horn, slightly paler yellow at base of lower mandible. Legs dark grey. Immature duller with scaly throat. **Voice** A sharp incisive *chick* or *spik*. Song a distinctive five-note (or longer) phrase, a single note followed by two couplets. First note quiet, next two higher and louder, and final two higher still and sharp. **SS** Unlikely to be confused. **Status** Shy resident; uncommon on Trinidad, rare on Tobago. Found above 200m in heavily forested areas with extensive leaf litter.

### Yellow-chinned Spinetail *Certhiaxis cinnamomeus cinnamomeus* 15cm

Adult has chestnut crown. Face grey-brown with diffuse pale supercilium and thin black lores extending slightly behind eye. Small yellow chin patch only visible at very close range. Throat white, rest of underparts off-white. Nape and mantle reddish-brown; wings, rump and tail bright chestnut, almost ginger. Bill and legs black. Juvenile recalls adult but duller above and flushed cream on sides and flanks. **Voice** Series of short dry, throaty rattles, and a *chup...chup* contact-call. **SS** Unlikely to be confused. **Status** Common and widespread Trinidad resident of wet meadows, freshwater marshes and mangrove edges. Absent from Tobago.

### Streaked Xenops *Xenops rutilans heterurus* 12.5cm

Adult has dark brown crown densely streaked buff, bordered by long, thin creamy white supercilium. Face brown with broad black eyeline and white malar. Throat white. Upper breast grey-brown with creamy teardrops; lower breast and belly olive-brown; undertail rufous. Nape and mantle warm chestnut-brown lightly streaked buff. Rump and tail more rufous. Flight feathers darker brown with buff bar at base. Bill short and upturned; dark horn above paler below. Legs black. Juvenile like adult. **Voice** Call a high-pitched thin *tzip*; song a rapid twittering. **SS** Unlikely to be confused. **Status** Uncommon resident of Trinidad forests. Usually climbs thin branches or saplings, often hanging upside-down. Absent from Tobago.

**Pale-breasted Spinetail**

juvenile

immature

adult

**Stripe-breasted Spinetail**

*carri*

*terrestris*

**Grey-throated Leaftosser**

adult

juvenile

**Yellow-chinned Spinetail**

adult

juvenile

**Streaked Xenops**

## WOODCREEPERS – SUBFAMILY DENDROCOLAPTINAE

Fifty-seven species occur throughout Neotropics. Locally, two species occur on both islands with a further two restricted to Trinidad and one found solely on Tobago. Rather uniform brown plumage and best separated by a combination of bill shape and colour, together with presence or absence of cream mottling on head and underparts.

### Plain-brown Woodcreeper *Dendrocincla fuliginosa meruloides* 22cm

Medium-sized and unstreaked. Adult has rich cinnamon-brown crown. Face grey with brown eyes. Underparts cinnamon-brown. Upperparts brown with wing-coverts and uppertail more rufous. Flight feathers dark grey. Bill very slightly decurved, dark horn above, paler below. Legs grey. Immature recalls adult. **Voice** Low-pitched, rather hollow trill comprising 20–25 notes, sometimes many more. **SS** From all other woodcreepers on Trinidad by lack of streaking on head and nape. On Tobago, from Olivaceous by larger size and browner plumage. **Status** Common Trinidad resident, favouring deciduous forest at all but highest elevations. Uncommon on Tobago.

### Olivaceous Woodcreeper *Sittasomus griseicapillus griseus* 15cm

Small, slim, small-billed and unstreaked. Adult has unmarked olive-grey head and underparts. Nape, mantle and wing-coverts olive-grey. Flight feathers grey with buff underwing-stripe noticeable in flight. Tertials, rump, uppertail-coverts and tail chestnut-brown. Bill short, slightly decurved and dark grey; legs grey. Immature recalls adult. **Voice** Thin, high-pitched trill, rising then falling and lasting c.4 seconds. **SS** Unlikely to be confused. **Status** Shy, uncommon resident of dry and wet forests on Tobago. Appears restless, quickly climbing trunk before flying to base of next tree. Absent on Trinidad.

### Straight-billed Woodcreeper *Xiphorhynchus picus altirostris* 23cm

Large and streaked with strong pale bill. Adult has dark grey-brown crown, nape and neck-sides densely streaked cream. Face and throat grey-white, with brown eyes. Upper breast brown with large tear-shaped cream spots; rest of underparts chocolate-brown. Upperparts brown; wings and tail more rufous. Bill substantial, straight and ivory-coloured; legs pale grey. Immature recalls adult. **Voice** Piercing loud trill starting quickly, then slowing. **SS** From Streak-headed Woodcreeper by larger size, all-rufous back and more robust, straight bill. **Status** Uncommon Trinidad resident, found in mangroves on E and W coasts. Absent on Tobago.

### Cocoa Woodcreeper *Xiphorhynchus susurrans susurrans* 23cm

Large and robust with long, stout, slightly decurved bill. Adult has olive-brown head, densely speckled buff, with diffuse black malar and dirty white throat. Breast olive-brown mottled buff, belly and undertail-coverts grey-brown. Upperparts olive-brown with dense cream mantle streaking, wings and tail rufous-brown. Bill black, legs grey. Immature recalls adult. **Voice** Series of full, rather melancholy notes; first 4–5 on same pitch, rest descending, and whole is monotonously repetitive. **SS** From Straight-billed and Streak-headed Woodcreepers by olive plumage and all-dark bill. **Status** Common and widespread resident in T&T.

### Streak-headed Woodcreeper *Lepidocolaptes souleyetii littorali* 20cm

Slight in structure, streaked, with thin, decurved bill. Adult has dark brown head densely streaked cream. Chin pale buff, rest of underparts buff-brown with cream streaking restricted to breast, sides and flanks. Upperparts olive-brown, wings and tail more rufous. Bill yellowish-pink, legs grey. Immature recalls adult. **Voice** Rather short, very high-pitched, fast and thin trill, *pee pee pee pee*. **SS** From Straight-billed Woodcreeper by smaller size, brown back and wing-coverts, and smaller decurved bill. **Status** Uncommon and local, in mangrove and coconut estates of S and SW Trinidad, occasionally in woodland of Aripo savanna and rare in E coast mangrove. Absent from Tobago.

Olivaceous Woodcreeper

Plain-brown Woodcreeper

Straight-billed Woodcreeper

Cocoa Woodcreeper

Streak-headed Woodcreeper

# ANTBIRDS – THAMNOPHILIDAE

Antshrikes are usually large-headed, short-winged, short-tailed and bulky, with heavy, hooked bills. Antvireos are small and dumpy with heavy bills, short wings and pot bellies. Antwrens are short-winged and even shorter tailed. Antbirds are long-billed, stout-legged and generally short-tailed, wary and skulking in habits, close to the forest floor.

## Great Antshrike *Taraba major semifasciatus*       20cm

Adult male has black head with shaggy crest and red eyes. Underparts white with grey wash to rear flanks. Undertail black with clear white tips. Upperparts black with white wing-covert tips and tail-sides. Female has chestnut-brown head, white underparts with buff rear flanks, chestnut-brown upperparts with darker flight feathers, and more rufous tail. Immature male recalls adult, with chestnut tones to wing-coverts and fine black scalloping below. Juvenile has brown eyes. **Voice** Series of notes (often 20+) on same pitch but increasing in speed, followed by a pause and cough *ker.ker.ker ker ... kiarh*; also a low-pitched harsh *churr churr churr*. **SS** None. **Status** Common Trinidad resident, in both dense scrub and forest. Absent Tobago.

## Black-crested Antshrike *Sakesphorus canadensis trinitatis*       17cm

Adult male has black shaggy-crested head. Throat and upper breast black, continuing as narrow centre to lower breast; breast- and neck-sides white, extending as nuchal collar. Rest of underparts grey. Mantle and rump chestnut-brown. Wings and tail black with white fringes and tips. Bill black, legs grey. Adult female has orange-brown shaggy crest, black-and-white cheeks and thin white orbital. Breast brownish-grey with faint dark streaking, nape and mantle tawny-brown, wing-coverts and tertials black, fringed and tipped white. Tail brown, fringed and tipped white. **Voice** Series of 8–10 notes increasing in speed and rising in pitch. A shrill, rapid *ke ke ke ke ke*. **SS** Unmistakable. **Status** Common Trinidad resident of lowland wet forest and mangrove. Absent from Tobago.

## Barred Antshrike *Thamnophilus doliatus fraterculus*       15cm

Adult male has black crown and bushy crest flecked white. Face greyish-silver with white eyes. Underparts white densely barred black. Upperparts silver-grey with dense black and white bars. Wings and tail black, narrowly barred white. Bill and legs blue-grey. Female has chestnut crown and crest; face grey streaked black. Throat pale creamy-buff, rest of underparts cinnamon. Upperparts chestnut. Immature male recalls adult, with varying amounts of buff on breast and belly, and reddish tones to crown and wing-coverts. **Voice** Song a harsh raucous laugh, accelerating and rising in pitch; also a single scolding *arrrh*. **SS** Unmistakable. **Status** Common and widespread resident in T&T.

## White-flanked Antwren *Myrmotherula axillaris axillaris*       11cm

Adult male has dark grey head. Throat and breast black, greyer on the belly. White flank patch is often partially obscured by folded wing. Undertail spotted white. Upperparts dark grey; tail black. Wings black with white covert spots. Stout bill and legs black. Female has grey-brown head. Throat buff-white, breast and belly warm tawny-buff, with creamy flanks. Upperparts warm brown with pale coverts spotting. Immature male slate grey where adult black, wings and tail as adult female. Juvenile like female. **Voice** A rather mournful nasal *eeuw*; song a fairly quiet 20–25-note descending trill. **SS** None. **Status** Common Trinidad resident in forest at all elevations Absent from Tobago.

## White-fringed Antwren *Formicivora grisea intermedia*       12.5cm

Adult male has greyish-brown head with broad white supercilium, bordering rear ear-coverts and extending to neck-sides and flanks. Rest of underparts black (race *tobagensis* on Tobago has white undertail-coverts). Upperparts warm brown; tail black tipped white. White dots on lesser and median coverts tips, and broad white bar on greater coverts. Female has brown head with creamy supercilium and greyish-yellow ear-coverts. Throat and breast pale grey to off-white on undertail-coverts, with dark streaks on breast/upper belly. Upperparts brown, wings black with two broken white wing-bars. Tail black. Juvenile warmer brown above than female, and yellowish-creamy below with fine streaks on upper breast. **Voice** A rich *chi-weh* and strident *hwip hwip* repeated frequently. Song a piping *peeer* followed by a tremulous trill. **SS** None. **Status** Common resident on Tobago and Chacachacare Is. Absent mainland Trinidad. Favours dry scrubby forest and forest edge.

Great Antshrike

♂

♀

♂
immature

Black-crested Antshrike

♂

♀

juvenile

Barred Antshrike

♂

juvenile

juvenile

♀

White-flanked Antwren

♂

♂
immature

♀

White-fringed Antwren

intermedia
(Trinidad)

♂

♀

juvenile

## Plain Antvireo *Dysithamnus mentalis andrei*      12.5cm

Adult male has dark grey head with black ear-coverts. Underparts pale grey, palest on throat, and sometimes tinged yellow. Upperparts dark grey with white tips to wing-coverts and outer fringes of scapulars. Female has dull chestnut-orange crown and grey face with thin white orbital ring. Throat white, breast and belly greyish-yellow, becoming very pale lemon on undertail-coverts. Upperparts olive with brighter green fringes to folded primaries; tail dusky. Bill and legs dark grey. Immature recalls respective adult, but duller and paler. **Voice** A rather melancholy *tchiow..tchiow* or *ciyup ciyup* repeated frequently. Song a series of 12–15 *churr* notes, first six on even pitch, rest descending quickly; also a 10-note descending 'bouncing-ball' trill. **SS** Unlikely to be confused. **Status** Uncommon forest resident in T&T.

## White-bellied Antbird *Myrmeciza longipes*      15cm

Adult male has orange-brown crown. Face black, bordered by long blue-grey stripe over eye, wrapping round ear-coverts and continuing on neck-sides and flanks. Throat and upper breast black. Rest of underparts white, tinged orange on rear flanks and undertail-coverts. Upperparts orange-brown. Bill black, legs strong and pale pink. Female duller with black restricted to ill-defined mask on otherwise grey face. Throat and breast dirty buff, belly and undertail-coverts white. Juvenile recalls adult female. **Voice** Shrill but liquid descending glissando of 18–20 notes, *teu teu teu*. A low-pitched loud, short trill, repeated frequently. **SS** Unlikely to be confused. **Status** Widespread yet uncommon Trinidad resident of deciduous woodland. Absent from Tobago. Skulking and unlikely to be seen without effort.

## Silvered Antbird *Sclateria naevia naevia*      15cm

Adult male has dark grey head. Throat and central upper breast white, faintly streaked black. Rest of underparts dark grey with tear-shaped white spots. Upperparts dark grey. Wings black with white spots on coverts tips; tail black. Bill black, legs pale yellowish-pink. Female has brown head. Underparts whitish with variable buffy-brown scallops. Upperparts brown but wings darker with variable buff spots on coverts; tail black. Juvenile recalls duller female with buffy terminal spots on wing-coverts. **Voice** Song a musical glissando which slowly rises then falls; call a hollow *tup*. **SS** Unlikely to be confused. **Status** Uncommon resident on Trinidad, but breeding unproven. Favours low-elevation forested streams and mangrove edges (diagnostic). Absent from Tobago.

# ANTTHRUSHES AND ANTPITTAS – FORMICARIIDAE

Of 12 species of antthrush and 51 antpittas in the Neotropics, only one of each occurs on Trinidad. Both are forest ground-dwellers and often initially detected by their vocalisations. Antthrushes are plump-bodied, short-legged and walk with a horizontal gait and their short tails held cocked. They often vocalise from a fallen log or tree stump. Antpittas are also plump, with an upright stance, and appear long-legged and almost tail-less. They forage by hopping through leaf litter, but occasionally sing from the midstorey.

## Black-faced Antthrush *Formicarius analis saturatus*      20cm

Adult has dark chestnut crown. Face black with blue-white orbital ring. Chin and throat black. Upper breast dark grey, paler below; undertail-coverts rufous. Nape chestnut, more rufous on neck-sides. Mantle and wings dark brown, brighter on rump and uppertail-coverts, tail black. Bill black, legs grey. Juvenile recalls dull adult, but lacks bright eye-ring and has faint white throat spots and buff flank bars. **Voice** Single loud whistled note followed by several others on descending scale; a two-note whistle, second lower pitched. Call a loud *chulik* or *tulip*. **SS** Unmistakable. **Status** Common Trinidad resident, favouring leaf-litter on forested slopes. Absent from Tobago.

## Scaled Antpitta *Grallaria guatimalensis aripoensis*      19cm

Adult has grey crown with extensive black scaling. Face grey with diffuse buff-white malar and thin white rear eye crescent. Throat brown, edged black with whitish lower border; rest of underparts warm reddish-orange. Upperparts olive-brown, scaled black, wings uniform brown. Bill and legs grey. Juvenile brown, with head and mantle scaling, buff wing-coverts spots and buff underparts streaking. **Voice** Song a hollow, tremulous trill, with last 8–10 notes louder and slightly higher pitched. **SS** Unmistakable. **Status** Rare resident of dense forest in Northern Range on Trinidad. Absent from Tobago. Not easily seen without effort.

**Plain Antvireo**

♂

♀

**White-bellied Antbird**

♂
immature

♂

♀

**Silvered Antbird**

♂

♀

♀

**Black-faced Antthrush**

juvenile

adult

**Scaled Antpitta**

juvenile

adult

# TYRANT FLYCATCHERS – TYRANNIDAE

A large, diverse, New World family of 437 species, of which 40 have been recorded in T&T. Many are resident, the rest migrants from North or South America. Most have predominantly green, yellow or brown plumage, many with conspicuous facial or crown markings and wing-bars.

### Crested Doradito *Pseudocolopteryx sclateri* 10cm

Adult has black, shaggy-crested crown with orange or yellow flecks. Face dark brown. Underparts yellow. Upperparts and tail olive-brown, wings black with two pale wing-bars and white tertial fringes. Bill and legs black. Immature duller with pale supercilium and buff underparts. **Voice** High-pitched, thin series of notes, *sik, tsweek, tsweee*. **SS** Unmistakable in marsh habitat. **Status** Very rare visitor, not reliably recorded in last ten years. Apparently once resident in freshwater marshes on Trinidad. Absent from Tobago.

### Olive-striped Flycatcher *Mionectes olivaceus venezuelensis* 12.5cm

Dark grey-green head with white postocular spot. Underparts lemon-yellow with dark olive streaking on breast and flanks. Upperparts bright olive-green; wings and tail fringed bright green. Bill dark with paler base. Legs pinkish-grey. Young browner on head, less green above, weaker yellow below. **Voice** Rarely heard; a high-pitched hissing *tsi tsi tsee*. **SS** Postocular spot and streaked breast render it unlikely to be confused. **Status** Confined to highest forest of Northern Range, Trinidad, usually in upper canopy. Absent from Tobago.

### Slaty-capped Flycatcher *Leptopogon superciliaris pariae* 12.5cm

Grey-white forehead and darker grey crown (darkest at rear). Face pale grey with large but ill-defined dusky ear-coverts spot. Underparts lemon-yellow, palest on throat and undertail-coverts. Mantle greyish lime-green; rump to tail grey with green outer webs to outer tail feathers. Wings grey-brown with two buff wing-bars. Tertials black, broadly fringed pale yellow or white. Bill dark grey, legs paler. **Voice** Disyllabic, rather flat-toned *wich-hu wich-hu* repeated regularly; also a harsh *chir*. **SS** None. **Status** Uncommon Trinidad forest resident, rare below 200m. Frequently seen on low perches beside forest trails. Absent from Tobago.

### Ochre-bellied Flycatcher *Mionectes oleaginous pallidiventris* 12.5cm

Olive-grey head. Underparts orange-buff with green upper-breast wash. Upperparts lime-green. Wings dark brown with two buff wing-bars and black tertials broadly fringed white. Upper mandible black, lower yellow. Legs black. **Voice** Sharp strident *sii-uu* or *see-ar*; several squeaky *tchik tchik* notes and series of quiet *chick* or *chew* notes. **SS** None. **Status** Uncommon forest resident in T&T; forages low, often near water.

### Spotted Tody-Flycatcher *Todirostrum maculatum amacurense* 10cm

Tiny. Adult has dark grey head, darkest on forecrown, with white loral spot. Eyes bright orange-yellow. Throat white, bright yellow below, streaked black from chin to breast and flanks. Nape dark grey, mantle and rump bright olive-green. Tail brown. Wings dark grey with green fringes to secondaries, yellow covert fringes and white tertial fringes. Bill disproportionately large, long and wide, mainly black, with pale underside. Legs brown. Immature has less streaking on upper breast, duller mantle and dark brown eyes. **Voice** Series of loud, strident *chee* and *chirp* notes. **SS** None. **Status** Locally common resident SW Trinidad, not found north to San Fernando. Almost always in trees near water; usually, but not exclusively, in mangrove. Absent from Tobago.

### Northern Scrub Flycatcher *Sublegatus arenarum glaber* 14.5cm

Slightly crested, small-billed and upright posture, recalling a miniature *Myiarchus* flycatcher Adult has dark brown crown. Face grey with thin white supercilium and black eyeline. Throat, breast grey, rest pale yellow. Upperparts olive-grey. Wings black with three pale wing-bars that reduce with wear. Tertials and tail black. Bill and legs dusky. Juvenile has whiter belly and pale fringes above. **Voice** A quiet whistled *pee-wee*. **SS** From *Myiarchus* by smaller size and tiny bill. Smaller Mouse-coloured Tyrannulet is round-headed and has a horizontal posture. **Status** Uncommon resident of mangrove and low-elevation dry forest on Trinidad and Bocas Is. Absent from Tobago.

### Short-tailed Pygmy Tyrant *Myiornis ecaudatus miserabilis* 7.5cm

Tiny and almost tail-less with very upright stance. Grey head, prominent white loral streak and orbital ring. Pale yellow flanks. Mantle and rump lime-green, tail dark brown. Wings black, fringed green. Bill thin and black, legs dark grey. **Voice** An often difficult to hear insect-like high-pitched buzzing trill, *tsee tsee tsee*. **SS** Size and shape render it unlikely to be confused. **Status** Scarce forest resident on Trinidad, rare above 400m. Easily overlooked as forages just below canopy. Absent from Tobago.

## Crested Doradito

## Olive-striped Flycatcher

## Slaty-capped Flycatcher

## Ochre-bellied Flycatcher

## Spotted Tody-Flycatcher

## Northern Scrub Flycatcher

juvenile        adult

## Short-tailed Pygmy Tyrant

## Forest Elaenia *Myopagis gaimardii trinitatis* 12.5cm
Adult has grey crown and nape, and obscure white central crown patch. Face mottled pale grey with broken white orbital, and black eye-stripe and thin white supercilium reaching just behind eye. Throat grey-white, rest of underparts lemon-yellow with faint streaking on upper breast. Mantle and rump olive-green, wing-coverts and tail olive-brown with two broad lemon-white wing-bars. Flight feathers dark brown fringed lime green, tertials fringed white. Thin bill black, paler at base of lower mandible. Legs dusky. Juvenile like adult. **Voice** Usual call is a 2–3 syllable *pit-swEET*, with the last two slurred; also musical, high-pitched jangling notes at dawn. **SS** From Yellow-olive Flycatcher by slender and smaller jizz, thinner bill, central crown-stripe, greater contrast between crown and mantle, olive upper-breast streaking and voice. From Slaty-capped Flycatcher by lack of ear-coverts markings and diffuse olive breast streaking. Forest Elaenia is small, active with a more horizontal stance. **Status** Fairly common Trinidad resident in forest and second growth. Absent from Tobago.

## Yellow-bellied Elaenia *Elaenia flavogaster flavogaster* 17cm
Untidy looking. Adult has brown crown with off-white central stripe, the feathers raised in a shaggy crest, parted in middle. Face greyish with whiter lores and eye-ring. Throat and upper breast grey, grading into dirty lemon central lower breast and belly; flanks grey. Upperparts olive-grey, wings and tail dark olive-brown with green sheen to flight feathers and two dirty white wing-bars. Tertials black, fringed white. Bill dusky above, pinkish-yellow below. Legs vinaceous grey. Juvenile recalls adult but lacks white in crown, and is overall paler. **Voice** Loud, scratchy *wheer du wheer du* and series of hoarse, *buzzy heer heer heer* notes. **SS** From Small-billed and Lesser Elaenias by larger size and more shaggy-crested appearance. From Southern Beardless Tyrannulet by much larger size, white crown patch and lack of supercilium and eyeline. **Status** Common and widespread resident in T&T.

## Slaty Elaenia *Elaenia strepera* 15cm
Adult male has dark grey head with obscure white central crown-stripe and thin yellow-white orbital. Throat and breast slate, belly and undertail-coverts whitish. Upperparts dark grey, wings and tail browner, fringed white, with two dirty white wing-bars. Upper mandible dusky, lower pinkish-yellow. Legs black. Adult female has yellow tone to belly, olive tone to upperparts and brighter rufous wing-bars. Juvenile male like adult, with cinnamon wing-bars and yellowish tone to belly. Juvenile female recalls adult. **Voice** Usually silent. **SS** None. **Status** Accidental to Trinidad, with one record in Northern Range, Jul 1998.

## Lesser Elaenia *Elaenia chiriquensis flavivertex* 13.5cm
Adult has olive-grey crown finely streaked black, and concealed white central stripe. Occasionally, rear crown can look ragged but not truly crested. Face olive-grey with thin white orbital ring. Chin and throat off-white, upper breast and flanks greyer. Lower breast and belly dirty greyish-yellow, paler on undertail-coverts. Upperparts olive-grey, wings and tail darker olive-brown with two, occasionally three, white wing-bars and white fringes to tertials. Upper mandible dark horn, lower dirty orange. Legs black. Juvenile recalls adult but has creamier underparts and three dirty wing-bars. **Voice** Harsh grating *dip waa* with second note lower pitched, a light *weep* and several raucous, typical *Elaenia* chatter notes. **SS** From Yellow-bellied Elaenia by smaller, slighter jizz, more rounded crown lacking obvious crest, less yellow on lower breast, and vocals. From extremely similar Small-billed Elaenia by dirtier underparts and vocals. **Status** Rare Trinidad resident, favouring lowland dry scrub. Absent from Tobago.

## Small-billed Elaenia *Elaenia parvirostris* 15cm
Clean-cut with rounded head, virtually no crest and upright jizz. Adult has olive crown with concealed white central stripe. Face olive with paler lores and distinct white orbital ring. Throat, breast and belly off-white with duskier sides, undertail-coverts yellowish. Upperparts olive, wings and tail dark brown with two obvious white wing-bars, a third indistinct lesser coverts bar and yellowish flight-feather fringes. Birds in worn plumage have significantly reduced white orbital and wing-bars. Bill does not appear smaller than other elaenias; dark grey above, pinkish-yellow below. Legs black. **Voice** Usually silent. **SS** Extremely difficult to separate from smaller Lesser Elaenia in field, but underparts cleaner and orbital more distinct. **Status** Accidental on Trinidad with just two old records in secondary scrub.

## Forest Elaenia

## Yellow-bellied Elaenia

worn
plumage

fresh
plumage

## Slaty Elaenia

♂

♀

## Lesser Elaenia

fresh
plumage

## Small-billed Elaenia

## Southern Beardless Tyrannulet *Camptostoma obsoletum venezuelae*    10cm

Tiny and shaggy-crested; tail often cocked. Adult has olive-brown crown; face olive-grey with ill-defined black eyeline and faint white orbital. Underparts pale lemon-yellow with olive wash to throat and upper breast. Upperparts grey-green, rump browner. Wings black with two white wing-bars, pale fringes to tertials and bright green flight-feather edges. Tail dark olive-grey. Bill grey with paler base to lower mandible; legs grey. Juvenile pale buff below, browner above with cinnamon wing-bars. **Voice** Many vocalisations, mostly on a minor scale, including a four-note call, first higher pitched, *clee chee chee chee*. **SS** From Yellow-bellied Elaenia by much smaller size. From Mouse-coloured Tyrannulet by smaller size, flatter crown and more olive plumage. **Status** Common resident on Trinidad, favouring forest, second growth and mangrove edge. Absent from Tobago.

## Mouse-coloured Tyrannulet *Phaeomyias murina incomta*    12cm

Distinctly flat-headed with horizontal jizz. Adult has grey-brown head and diffuse supercilium. Throat and upper breast grey, belly and undertail-coverts pale lemon-yellow. Upperparts grey-brown; rump slightly paler, with two well-defined, dull wing-bars. Bill dark above, pinkish-orange below; legs grey. Juvenile has speckled crown, drabber upper- and underparts, with three off-white wing-bars. **Voice** Thin, slightly disyllabic *su-eet* or *chu-wee* and a quiet rasping chatter. **SS** From similar Northern Scrub Flycatcher by posture, more robust bill with paler lower mandible, greyer crown, diffuse supercilium and lack of black eyeline. **Status** Locally common resident of dry scrub forest on Bocas Is, and scarce visitor to mainland Trinidad, favouring dry scrubby hillsides and mangrove edge. Absent from Tobago.

## Yellow-olive Flycatcher *Tolmomyias sulphurescens berlepschi*    14cm

Adult has olive-grey head, white pre-loral spot and thin supercilium extending just over eye. Eyes and eye-ring usually pale but this well-documented feature is often difficult to see in field. Throat pale grey, breast dull olive, belly and undertail-coverts dull lemon-yellow. Nape grey, mantle and rump olive-green. Wings and tail dusky, fringed pale, with two yellowish wing-bars. Bill broad and flat; upper mandible black, lower pink. Legs black. Juvenile recalls adult but paler, dull sulphur-yellow below with dark eyes. **Voice** Call, *chip chip*. **SS** From smaller Yellow-breasted Flycatcher by smaller head, greyer face and duller underparts. From Forest Elaenia by lack of white crown-stripe and black loral line, obscure supercilium and lack of breast streaking. **Status** Uncommon Trinidad forest resident. Absent from Tobago.

## Yellow-breasted Flycatcher *Tolmomyias flaviventris collingwoodi*    12.5cm

Small, plump and large-headed. Adult has bright yellowish-green head with ochre lores, black eyes and thin yellow eye-ring. Underparts bright yellow. Upperparts bright yellowish-olive, tail brown. Wings black fringed lime-green, with two yellow wing-bars. Bill black and rather flat with pale underside, legs vinous grey. **Voice** Disyllabic *soo it* with emphasis on first note, second higher pitched, reminiscent of Forest Elaenia. **SS** From Yellow-olive Flycatcher by smaller size, plain yellowish-green face, lack of white eye markings and bright yellow underparts. **Status** Common resident in T&T, found in forest, second growth, cultivated estates and mangrove edge.

## White-throated Spadebill *Platyrinchus mystaceus insularis*    10cm

Small, dumpy, large-headed and almost tail-less. Adult has olive-brown head, tawny-yellow supercilium and broad orbital ring; two vertical black bars traverse cheeks and ear-coverts, with a tawny patch between them, and indistinct tawny loral spot. Yellow crown patch (lacking in female) rarely visible in field. Chin and throat creamy white, rest of underparts yellow-buff. Mantle and rump olive-brown, wings and tail darker with pale fringes to coverts and tertials. Bill dark horn above, paler below and very broad-based. Legs pink. Immature recalls adult. **Voice** Song is a long, fast trill that rises, then falls. Call is a quiet tinny *kwik* or *squik* and doubled *wipput wipput*. **SS** None. **Status** Uncommon forest resident, widespread on Trinidad, but mainly on Main Ridge on Tobago. Usually perches low to ground, remaining motionless for long periods, before flying fast and direct to the next perch.

**Southern Beardless Tyrannulet**

juvenile

adult

**Mouse-coloured Tyrannulet**

**Yellow-olive Flycatcher**

**Yellow-breasted Flycatcher**

**White-throated Spadebill**

♂

♀

## Bran-coloured Flycatcher *Myiophobus fasciatus fasciatus* 12.5cm

Small and round-headed. Adult has reddish-brown crown with obscure central yellow patch. Face grey-brown with faint buff supercilium and subdued eye-stripe. Underparts pale yellow or off-white with variable, dense but thin dark upper-breast streaking. Upperparts bright reddish-brown, wings much darker with orange-buff tips to wing-coverts and tertials. Bill black above, yellow below. Legs black. Juvenile more rufous above with cinnamon wing-bars and lacks central crown patch. **Voice** Low-pitched, ascending trill. **SS** None. **Status** Uncommon Trinidad resident, more numerous on Bocas Is. Favours lowland savannas, secondary growth and arid hillside scrub. Absent from Tobago.

## Olive-sided Flycatcher *Contopus cooperi* 16cm

Dull-coloured, bull-headed and stocky, with stout bill, sloping forehead and short, slightly notched tail. Adult has plain olive-grey head. Chin and throat white, neck-sides, breast and belly olive-grey. Central underparts creamy white, narrow on upper breast, broader on belly and undertail-coverts. Upperparts olive-grey, wings and tail darker, with two indistinct pale wing-bars and white tertial fringes. Bill black with yellow-orange base to lower mandible; legs black. Juvenile recalls adult with buff wing-bars and white rump-sides. **Voice** Trisyllabic *per di-eu*. **SS** From smaller Tropical Pewee by more robust shape, longer wings, bulkier bill, more clearly defined white central underparts stripe, and duller wing-bars. **Status** Uncommon visitor to Trinidad, favouring forest clearings above 300 m. Frequently perches on utility wires or exposed branches. Mostly late Sep–early Apr. No records from Tobago.

## Tropical Pewee *Contopus cinereus bogotensis* 14cm

Small, upright, with peaked crown and long tail. Adult has olive-grey crown and darker forecrown. Face grey with whitish lores and chin. Breast and belly dirty olive-grey with pale yellow central stripe. Upperparts olive-grey, wings darker with two wing-bars and pale fringes. Tail olive-brown and notched, rather than forked. Upper mandible dark horn, lower bright orange. Legs black. Juvenile has paler lemon-grey underparts, pale grey nape and mantle, blacker wings with broad buff coverts and tertial fringes. **Voice** Shrill *shreee*, reminiscent of referee's whistle. **SS** From larger Olive-sided Flycatcher by conical head, longer tail, smaller bill, yellower central underparts and better defined white wing-feather fringes. From Euler's Flycatcher by more rounded head, lack of orbital ring and whiter wing-bars **Status** Common Trinidad resident, favours forest and cultivated estates. Absent from Tobago.

## Fuscous Flycatcher *Cnemotriccus fuscatus cabanisi* 16cm

Adult has grey-brown head, distinct creamy supraloral (slightly broader behind eye), and black loral and eye streaks. Underparts pale lemon, sometimes creamy, with grey wash to neck-sides and on upper breast. Upperparts grey-brown, tail darker. Wings sooty-brown with two buff wing-bars and flight-feather fringes. Bill and legs black. Juvenile recalls adult. An olive colour morph exists, but no documented records in T&T. **Voice** Sweet *swit swit* or *sweep sweep*. **SS** Lack of olive or green plumage tones eliminate all other flycatchers. **Status** Uncommon Trinidad resident, mainly at mid elevations and usually in dry forest. On Tobago, common in both wet forest and dry second growth. Tends to forage from perches close to ground.

## Euler's Flycatcher *Lathrotriccus euleri lawrencei* 13.5cm

Small, drab and round-headed. Adult has dull olive-brown head, thin white orbital ring and faint white loral stripe. Chin and throat pale grading into dusky olive-brown breast. Belly and undertail-coverts dirty lemon-yellow. Upperparts olive-brown, wings and tail darker with two buff or dirty white wing-bars and white-fringed tertials. Bill dark grey above, dirty yellow below. Legs black. Juvenile recalls adult. **Voice** Plaintive, descending series of 4–6 notes, repeated every c.10 seconds with emphasis on first note, *peeer pi he he he*. **SS** From Tropical Pewee by rounded not conical head, white orbital and less distinct wing-bars. From Mouse-coloured Tyrannulet by range and habitat, more olive plumage, orbital and lack of supercilium. **Status** Uncommon Trinidad forest resident, sometimes shy. Tends to perch in lower and midstoreys, often in bamboo and quivers its tail while singing. Recent undocumented reports from Tobago.

## Bran-coloured Flycatcher

juvenile

adult

variation in
head colouring

## Olive-sided Flycatcher

## Tropical Pewee

## Fuscous Flycatcher

olive
morph

grey
morph

## Euler's Flycatcher

## Great Kiskadee *Pitangus sulphuratus trinitatis* 23cm

Large, robust and heavy-billed. Adult has black head with broad white stripes that meet on nape, and thin yellow central crown-stripe. Chin/throat white, below bright yellow. Upperparts brown with bright rusty fringes to coverts, flight and tail feathers. Bill and legs black. Juvenile has broader rufous fringes above and lacks yellow crown-stripe. **Voice** Single high-pitched strident *keer*; the familiar *ki-ka-deer* that gives rise to its name, and a mournful *kee wah*. **SS** From Boat-billed by leaner shape, smaller, slighter bill, richer brown mantle, rusty fringes to wings and tail, and supercilia meeting on nape. **Status** Abundant, widespread resident on Trinidad. Introduced but subsequently extirpated on Tobago.

## Piratic Flycatcher *Legatus leucophaius leucophaius* 15cm

Rather small with tiny bill, often initially located by far-carrying call. Adult has dark grey-brown head with obscure yellow central patch, ill-defined dirty white supercilia which nearly meet on nape, black loral line reaching just behind eye, broad white moustachial and thin black malar. Throat and upper breast dirty grey-white, diffusely streaked olive, with distinct lemon tone to flanks; rest yellowish. Upperparts dull brown with thin, pale fringes to coverts and tertials, variable rufous fringes to uppertail-coverts. Bill and legs black. Juvenile lacks yellow on crown, has paler yellow underparts, is browner with cinnamon wing-bars. **Voice** Oft-repeated, mournful *swee-uu*, occasionally *pi pi pid sweeuu*. **SS** From Variegated by smaller size, shorter all-dark bill, shorter tail, less rufous in uppertail-coverts and less contrast in greater coverts. From Streaked by much smaller size, smaller bill, less defined underparts streaking, and darker and duller plumage. **Status** Common breeding visitor to Trinidad, mostly in Feb–Oct, with some year-round. Uncommon on Tobago. Favours open scrub, cultivated estates and hillsides, seeking out nests of orioles, oropendolas and caciques.

## Boat-billed Flycatcher *Megarhynchus pitangua pitangua* 23cm

Large and bulky with disproportionately bulbous bill. Sooty-black head with broad white stripes from base of bill to rear crown, not quite meeting on nape. Yellow central crown-stripe sometimes obscured. Chin and throat white, rest of underparts rich yellow. Upperparts olive-brown with grey-white wing-feather fringes and narrow rufous tail fringes. Bill dark grey, extremely broad and large. Legs black. Juvenile lacks yellow crown patch and is paler below. **Voice** Harsh drawn-out *krr krr reeek*; a short shrill *chiru* or *chirilu*, and rapid four-note *churr-churr-churr*. **SS** See Great Kiskadee. **Status** Common and widespread Trinidad resident, rare above 600m. Absent from Tobago.

## Streaked Flycatcher *Myiodynastes maculatus tobagensis* 23cm

Large and powerfully built with long tail and rather short wings. Chestnut-brown crown, densely streaked black, and obscure yellow central patch. Face brown with long creamy supercilium, blackish mask from bill to behind eye, creamy white lower cheeks and black moustachial. Throat white and lightly streaked. Breast and belly off-white, streaked black on upper breast; undertail-coverts cream, lemon or occasionally pale orange. Upperparts warm brown mottled black, becoming chestnut-orange on rump and tail. Wing-coverts and tertials black, fringed pale grey; flight feathers dark brown, fringed golden-brown. Heavy bill, black above, pale lower base. Legs dark grey. Young birds lack yellow central crown. **Voice** Series of rather strident *chip* notes. **SS** From Piratic and Variegated Flycatchers by much larger size, bolder streaking and orange tail. **Status** Common forest resident in T&T.

## Variegated Flycatcher *Empidonomus varius varius* 18cm

Medium-sized and heavily streaked. Dark grey-brown crown, concealed yellow central patch, broad white supercilia that meet on nape, black mask from bill to ear-coverts, bold white moustachial and black wedge-shaped malar which does not reach bill. Throat white, grading to pale sulphur below, with dusky streaked breast and sides, dark brown above with bold white fringes to wings; uppertail-coverts and tail dusky, with broad rufous fringes. Small dark bill, pale at base. Austral migrant race *rufinus*, smaller, paler brown above and far less streaked below. Immature lacks yellow crown patch. **Voice** High-pitched, very thin *zweee*. **SS** From Piratic by larger size, longer two-toned bill, longer tail, more clearly defined greater coverts fringes, and more rufous on rump and uppertail-coverts. From Streaked by smaller size and structure, less distinct face and other markings. **Status** Rare non-breeding visitor to forest edge on Trinidad, *varius* from mainland, *rufinus* from south of continent, mostly Jan–Aug. Absent on Tobago.

Piratic Flycatcher

Great Kiskadee

adult

juvenile

Boat-billed Flycatcher

adult

juvenile

Streaked Flycatcher

Variegated Flycatcher

varius

rufinus

## Grey Kingbird *Tyrannus dominicensis vorax* 23cm

Robust, with short wings and long, slightly forked tail. Adult has grey head, concealed orange central patch and darker mask. Underparts white with pale grey wash to breast. Back grey, rest of upperparts darker grey-brown with pale wing-feather fringes. Heavy bill and legs black. Immature similar with rufous fringes to wing-coverts and tail feathers. **Voice** High-pitched chatter, distinctly drier and harsher than Tropical Kingbird. **SS** None. **Status** Locally uncommon resident in W and NE Trinidad, favouring open fields, freshwater marshes and mangrove edges. Much commoner on Tobago.

## Tropical Kingbird *Tyrannus melancholicus satrapa* 22cm

Large and sleek with heavy bill and well-forked tail. Adult has grey head with concealed orange central patch and ill-defined dusky eyeline. Throat white, upper breast pale lime-yellow becoming bright yellow on lower breast and belly. Nape powder grey, grading into grey-green mantle with brighter lime-green rump mottled black. Tail black. Wings dark olive-grey with pale fringes to coverts, scapulars and tertials. Bill and legs black. Immature paler and cleaner looking with buff fringes to wing feathers and uppertail-coverts. **Voice** Short, high-pitched 4–5-note trill. **SS** From Sulphury Flycatcher by forked tail, larger head and longer, thinner bill. **Status** Abundant resident in T&T, often seen on utility wires.

## Swainson's Flycatcher *Myiarchus swainsoni swainsoni* 20cm

Throat to breast greyish, very pale yellow to undertail-coverts. Olive-brown head and upperparts, with two whitish wing-bars. Bill pink to orange, dark tip above. Legs black. Juvenile has narrow rufous fringes to flight feathers. **Voice** Melancholy *phweee*. **SS** Some adults separable from Venezuelan Flycatcher by overall paler appearance, lack of rufous fringes to wings and tail, more extensive orange on bill and much paler yellow underparts. Many inseparable in field. **Status** Accidental to Trinidad, not recorded since 1958.

## Venezuelan Flycatcher *Myiarchus venezuelensis* 20cm

Sleek and rather thin-billed. Adult has uniform dark grey-brown head. Throat and upper breast very pale blue-grey, rest of underparts lemon-yellow, undertail pale olive. Upperparts dark grey-brown with obscure rufous fringes to outer webs of tail feathers. Wings browner with two white wing-bars, dull rufous fringes to flight feathers and pale tertial fringes. Bill thin and black, legs dark grey. Immature has slightly broader rufous fringes to primaries and uppertail-coverts. **Voice** A thin, mournful, and drawn-out whistle. **SS** From Brown-crested Flycatcher by slimmer, slighter all-dark bill, darker grey-brown crown and rufous restricted to fringes of tail feathers. Best distinguished by call. **Status** Scarce but fairly widespread resident in hill forest of C and NE Tobago. No acceptable records for Trinidad.

## Brown-crested Flycatcher *Myiarchus tyrannulus tyrannulus* 22cm

Large, robust, long-tailed and bushy-crested. Adult has warm brown crown contrasting with plain grey-brown face and reddish-brown eyes. Throat and upper breast pale blue-grey, rest of underparts lemon-yellow. Underside of central tail feathers rufous, rest darker and greyish. Upperparts grey-brown, wings dark brown with wing-coverts and tertials broadly fringed white. In fresh plumage, primaries fringed rufous and secondaries lime-yellow. Uppertail dark brown with obscure rufous inner webs. Bill black with pale pinkish-orange base to lower mandible. Legs dark grey. Immature has fringes of wing-coverts tawny and tertials edged buff. **Voice** A single *plik* repeated regularly, and sharp *whit*. **SS** From Venezuelan Flycatcher by more robust bill with pale base to lower mandible and extensive rufous on underside of central tail feathers. Otherwise, only safely separated by vocals. **Status** Common Tobago resident, less widespread on Trinidad. Favours lowland forest and mangrove.

## Dusky-capped Flycatcher *Myiarchus tuberculifer tuberculifer* 18cm

Peaked crown and delicate structure. Sooty-grey head, darkest on forecrown. Chin, throat and upper breast grey-white, rest of underparts pale lemon-yellow. Back olive-grey; wings darker with two indistinct, buff-white wing-bars and olive-brown panel to secondaries. Rump/uppertail-coverts chestnut-brown, tail dark brown. Bill black and rather fine, legs dark grey. Immature has rufous fringes to flight and tail feathers. **Voice** Mournful, disyllabic *sur-eep* or *hwii-uu*. **SS** From other *Myiarchus* by smaller size, much darker head, slighter bill, indistinct pale wing markings and rufous restricted to tail base. **Status** Uncommon Trinidad resident of forest edge, usually above 300m. Absent Tobago.

Grey Kingbird

Tropical Kingbird

Venezuelan Flycatcher

Swainson's Flycatcher

Dusky-capped Flycatcher

Brown-crested Flycatcher

## Fork-tailed Flycatcher *Tyrannus savana savana* 31–40cm

Adult male has sooty-black crown to below eye level with very thin orange central crown-stripe. Lower face, half-collar and underparts white. Lower nape and mantle grey; rump black. Tail extremely long, graduated and deeply forked, black with white on outer feathers. Wings dark grey-brown with paler fringes to coverts and tertials. Bill and legs black. Female like male but slightly shorter tail. Juvenile duller and browner with buff fringes to wing- and tail-coverts. **Voice** Hard, short, sharp *tchit!* **SS** None. **Status** Very common visitor to both islands, Apr–Oct. Some remain year-round. Breeding unconfirmed. Frequents open savanna, freshwater marsh and coastal scrub, roosting in very large numbers. Austral migrant race, *T. s. monachus*, occurs in small numbers, and is paler grey above with a complete white collar.

## Sulphury Flycatcher *Tyrannopsis sulphurea* 19cm

Stocky and plain-faced with rounded head and square-ended tail. Adult has grey crown with concealed yellow central patch and dark, dirty-looking face. Chin and throat white. Upper breast olive-yellow, rest of underparts yellow. Upperparts olive-brown. Wings brown with faint buff covert fringes. Bill and legs dark grey. Juvenile recalls adult. **Voice** Series of squeaks and squeals, *jür peep* and *squïi prrr*; also a loud, long drawn-out *sweeep*. **SS** From Tropical Kingbird by smaller, dumpier shape, generally dirtier looking plumage, shorter bill and square tail. **Status** Locally common resident of savanna with Moriche palms in C and E Trinidad. Absent from Tobago.

## Bright-rumped Attila *Attila spadiceus spadiceus* 20cm

Large-headed, chunky and rather short-tailed. Adult has greyish-olive head with reddish eyes. Throat and breast pale grey with variable, sometimes dense, diffuse olive streaking. Belly and undertail-coverts lemon-yellow, occasionally white. Nape and mantle olive-brown. Large square rump patch orange or yellow (usually only seen in flight). Tail rufous-brown. Wings dark brownish-grey with two indistinct buff wing-bars. Bill strong, stout and decidedly hooked; brown with darker tip. Legs dark grey. Juvenile recalls adult. Rufous morph, infrequently seen on Trinidad, has olive tones replaced by rufous-brown and much brighter tail. **Voice** A loud, seven-note, ascending *chulu chulu*, a sharp, short rattle and sweet *clui clui clui*. **SS** None. **Status** Scarce resident of forested slopes on Trinidad, keeping to denser foliage and usually detected by voice. Absent from Tobago.

**Fork-tailed Flycatcher**

juvenile

adult

variable length

**Sulphury Flycatcher**

**Bright-rumped Attila**

rufous morph

typical morph

### Pied Water Tyrant *Fluvicola pica pica*     15cm

Chunky with horizontal gait. Adult white with black rear crown, nape, fore mantle and wings. Tertials black with broad white fringes, and black tail has white terminal band. Bill and legs steel grey. Immature duller, less clean-cut with black parts of adult dirty brown. **Voice** A dry low-pitched disyllabic *dwi eeu*, and very short *typ* or *tip*. **SS** None. **Status** Common resident of freshwater marshes on Trinidad. Tends to fly very low and regularly alights on ground. Just one, undocumented, sighting from Tobago, Jul 2003.

### White-headed Marsh Tyrant *Arundinicola leucocephala*     12.5cm

Upright stance, rounded head and short wings. Adult male has white crown, face, chin and throat, with black eyes. All other body feathers black; wings and tail dark grey. Upper mandible black, lower bright orange-yellow with black tip. Legs black. Female appears pallid: head and underparts white with grey wash to breast and belly, upperparts soft grey with darker brown cast to wings and sooty tail. Juvenile like female. **Voice** Occasionally gives a high-pitched thin *tsip*. **SS** None. **Status** Common Trinidad resident of freshwater marshes and mangroves. Absent from Tobago.

## TITYRAS AND BECARDS – TITYRIDAE

Recent studies have shown that *Tityra* and *Pachyramphus* (and other related genera) should be placed in a new family, Tityridae. They were previously placed in Cotingidae or Tyrannidae, and have even been shown to be related to Manakins (Pipridae). The two species in Trinidad & Tobago are typically birds of the canopy, usually found singly or in pairs.

### Black-tailed Tityra *Tityra cayana cayana*     23cm

Adult male has black hood, with bright red orbital and loral skin. Underparts white. Mantle, wing-coverts and rump pale blue-grey, appearing white, flight and tail feathers black. Bill red with black tip, legs grey. Female has duller red lores and coarse black streaking on upper breast and back. Juvenile recalls adult female but orbital skin pink and lacks red on bill; much less streaking on breast. **Voice** Hoarse, amphibian-like croaking. **SS** None. **Status** Uncommon Trinidad resident of forest edge, savanna, cultivated estates and open scrubby hillsides. Absent from Tobago.

### White-winged Becard *Pachyramphus polychopterus tristis*     15cm

Heavy-set with large head and graduated tail. Adult male has black head. Rest of underparts dark grey with two rows of large white spots on undertail. Upperparts sooty-black with jagged white stripes formed by fringes to scapulars and wing-coverts. Tail black with outer feathers fringed white. Bill and legs dark grey. Female has grey-brown head with bold white loral stripe grading into broken orbital ring. Lower face and upper breast paler olive-grey, breast and belly pale olive-yellow. Undertail has large cinnamon terminal spots. Upperparts olive-brown. Tail black, outer feathers tipped buff. Wings black with two broad cinnamon wing-bars and fringes to flight feathers. Juvenile female resembles a duller adult, juvenile male goes through various phases to reach full immature (see plate). **Voice** A rapid *tiew tiew tiew* and varied sweet warbling and chattering notes. **SS** Adult male from male White-shouldered Tanager by larger size, undertail markings and white wing-coverts stripes. Female unlikely to be confused. **Status** Scarce resident in T&T of secondary forest and mangrove edge.

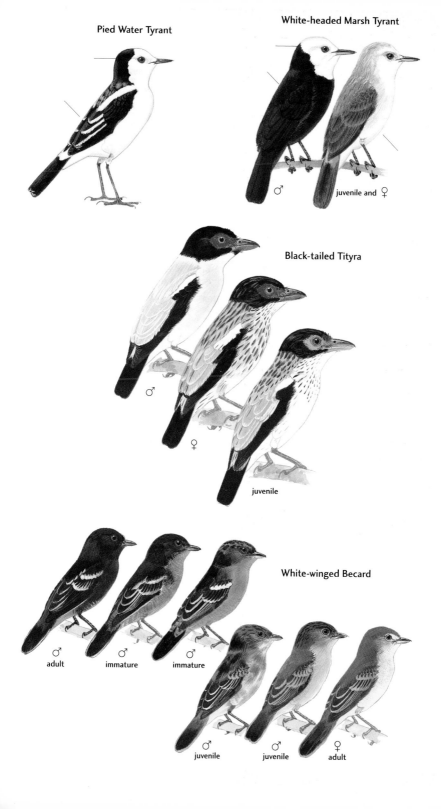

Pied Water Tyrant

White-headed Marsh Tyrant

♂     juvenile and ♀

Black-tailed Tityra

♂

♀

juvenile

White-winged Becard

adult     immature ♂     immature ♂

juvenile ♂     juvenile ♂     adult ♀

# COTINGAS – COTINGIDAE

Stocky with rounded wings, large heads and broad bills, Adult males are generally located by vocalisation. Despite their bright plumage, they are often well concealed in the midstorey canopy. Female and immatures are shy and seldom seen.

### White Bellbird *Procnias albus albus*                                                          30cm

Adult male entirely white with single black wattle from base of bill. Bill and legs blackish. Female has olive head. Underparts lemon-yellow, coarsely streaked olive. Upperparts dark olive-green. Young like adult female. **Voice** Much more musical than Bearded Bellbird. A two-note chiming *ding ding* or *un tinn*, the first note lower pitched. **SS** See Bearded Bellbird; adult male unmistakable. **Status** Very rare visitor to Northern Range, Trinidad, with only six records of males, all in Apr–Sep. None from Tobago.

### Bearded Bellbird *Procnias averano carnobarba*                                          28cm

Adult male has rich chocolate-brown hood. Throat has black, stringy, wattles up to several inches long. Underparts white. Back, rump and tail white; wings black. Bill and legs black. Female has olive-grey head, grey ear-coverts, olive throat streaked grey and greenish-yellow breast and belly densely streaked olive. Upperparts dark olive-green. Juvenile like female, but clearly streaked yellow on olive throat and breast, and streaked above. Male has several distinct phases before attaining basic adult in fourth year (see plate). **Voice** Far-carrying and often ventriloquial clunk repeated every few seconds. A series of quicker notes, slowly rising in pitch, *tonk-tonk-tonk*. **SS** Adult male unmistakable. Female and immature with care from rare White Bellbird by olive head and throat streaking. **Status** Locally common resident in Northern Range, Trinidad, favouring deciduous forest at 200–600m. Heard much more often than seen, but males stop calling mid-Aug–Oct. Absent from Tobago.

# MANAKINS – PIPRIDAE

Small, dumpy, large-headed, short-winged, mainly short-tailed and small-billed. Adult males brightly coloured and unmistakable. Females are dressed in shades of green and best separated by bare-parts colorations. Noted for their elaborate courtship dancing and displays.

### White-bearded Manakin *Manacus manacus trinitatis*                                  11.5cm

Adult male has crown to nape black, broad collar and entire underparts to undertail-coverts pure white, upper- and undertail-coverts and thighs pale grey; back, wings and tail black, glossed deep blue. Bill black, legs bright orange. Female/immature male uniform dark olive-green with distinct greyish cast. **Voice** Call a rasping *cheer* or *cheeow*. At leks, adult males utter a double *whit-whip* and make a snapping sound by flicking their secondaries. **SS** Adult male unmistakable. Female from Golden-headed Manakin by olive-grey plumage, orange legs and dark bill. **Status** Common Trinidad forest resident. Perches lower than other manakins. Absent from Tobago.

### Blue-backed Manakin *Chiroxiphia pareola atlantica*                                    14cm

Adult male has bright red crown with slight rear crest. Face, neck collar and underparts black. Mantle and wing-coverts pale blue with black rump, flight feathers and tail. Bill black, legs reddish-orange or yellow-orange. Female drab olive-green with slightly paler, more buffy-yellow underparts. Young male gradually acquires blue, red and black. **Voice** Loud and far-reaching, *chup*, *che-weep hu-weep* and *hweep hwuu*. At leks, males dance with whirring *naaaarrrr*, repeated regularly. **SS** None. **Status** Locally common resident of Tobago's forests. Recent unconfirmed reports from Trinidad.

### Golden-headed Manakin *Pipra erythrocephala erythrocephala*                    9cm

Adult male has bright golden-yellow hood, and milky-white eyes. Nape band red, rest of body feathers black. Bill pinkish-grey, legs pinkish-yellow with obscure red thighs. Female and juvenile male olive-green tinged yellow with dark eyes. **Voice** Sweet high-pitched *tieeu*, *tieeu*, *chew* and varied bubbles and trills. Also *triiow*, recalling White-bearded but somewhat sweeter. **SS** Adult male unmistakable; female from White-bearded Manakin by yellower plumage tones and different bare-parts colour. **Status** Very common and widespread resident of woodland and forest on Trinidad. Absent from Tobago.

White Bellbird

Bearded Bellbird

♀

♂

♂
third
year

♂
fourth
year

juvenile

♀
adult

♂
first
year

♂
second
year

White-bearded Manakin

♂

♀

Blue-backed Manakin

♂
juvenile

♂
immature

♀

Golden-headed Manakin

♂

♂
immature

♂
immature

♀
(and juvenile)

# PLATE 85: VIREOS AND GREENLETS

## VIREOS – VIREONIDAE

Most vireos are highly migratory; all are rather plain, some with bold face markings and wing-bars, recalling chunky warblers with heavier bills. Greenlets are visually dull and small but aggressive vocalisers. Just one species of peppershrike occurs, on Trinidad.

### Rufous-browed Peppershrike *Cyclarhis gujanensis flavipectus*                17cm

Large-headed, short-winged and dumpy. Grey head and nape with broad, sweeping, brick red eyebrow from bill to rear ear-coverts, and orange eyes. Chin white, throat and breast lime-yellow, rest of underparts grey-white. Upperparts grey-green. Stout, hooked bill and legs pinkish-grey. Young have buff wing-covert tips. **Voice** Musical with wide variety of song snatches, consistent tonal quality and 5–6 notes. Alarm a mournful wren-like scolding. **SS** Unmistakable. **Status** Common and widespread Trinidad resident. Absent Tobago.

### White-eyed Vireo *Vireo griseus* (race unknown)                12.5cm

Grey head with yellow lores and orbital, white eyes, and thin black eyeline. Throat white, rest of underparts pale grey with yellow-green wash to flanks and upper belly. Nape grey, mantle, rump and tail olive-green. Wings black with two white wing-bars; white tertial fringes and lime-coloured flight-feather fringes. Bill and legs dark grey. Young have dark eyes. **Voice** Usually silent. **SS** None. **Status** Accidental, Buccoo Marsh, Tobago, Jan 1998.

### Yellow-throated Vireo *Vireo flavifrons*                12.5cm

Bright olive-green head with bold yellow spectacles and lores, bordered black. Throat and breast bright yellow; rest of underparts white. Back olive-green, lower back to tail grey-green. Wings black with two white wing-bars and flight-feather fringes. Bill and legs dark grey. **Voice** A rapid, harsh *shep* and descending phrase of 3–5 harsh, grating notes. **SS** None. **Status** Accidental in T&T. Just one record in last 30 years, on Tobago, Jan 2005.

### Red-eyed Vireo *Vireo olivaceus*                14cm

Generally olive-green above and white below. Head grey with white eyebrow, bordered black above, and grey eye-stripe below; variable wash yellow below, eyes red, bill two-tone, black and grey, legs grey. Two resident races are mostly white below; *chivi* from the south is comparatively dull and dingy, whilst northern *olivaceus* is more like residents. Young are duller and have brown eyes. **Voice** Residents sing two-syllable, full, fruity *chi-woo*, much like Golden-fronted Greenlet. Migrants silent. **SS** From Black-whiskered by absence of malar stripe and usually cleaner, brighter plumage. **Status** Deciduous woodland and cultivated estates, usually at lower elevations.

### Black-whiskered Vireo *Vireo altiloquus*                15cm

Generally very much like Red-eyed Vireo, but separated by black submalar line, which is not always clear or apparent. Young duller, tending to buffy below and have brown moustachial. Race *barbadensis* is noticeably greener, and the two passage migrants duller and darker. Race *altiloquus* not documented for T&T. **Voice** Usually silent. **SS** From Red-eyed by black malar and stronger bill. **Status** Scarce visitor to forested areas of Trinidad; just two records on Tobago. Most sightings Nov–Mar.

### Golden-fronted Greenlet *Hylophilus aurantiifrons saturatus*                12.5cm

Olive-green above with golden-brown forehead and crown, and buffy cheeks; pale whitish below with ochraceous breast and flanks, pale yellowish undertail-coverts. Bill dark above, pinkish below; legs grey. Young pale sulphur-yellow below, richer on undertail-coverts. **Voice** A scolding *chur chur* or *dzer dzer*. Song recalls Red-eyed Vireo but thinner and trisyllabic, *chee weeooo*. **SS** None. **Status** Very common resident of forest, estates and second growth throughout Trinidad. Absent from Tobago.

### Scrub Greenlet *Hylophilus flavipes insularis*                12.5cm

Olive above, brownish on crown, with lime-green fringes to tertials and secondaries. Buffy below, with yellowish undertail-coverts. Bill dark above, pale below; legs grey; eyes usually brown. Juvenile has paler eyebrow. **Voice** A harsh nasal *zeer zeer*; song a musical series of 10–20 notes on even pitch, *tree tree tree tree*. **SS** None. **Status** Common resident on Tobago, favouring scrub and forest edge. Absent from Trinidad.

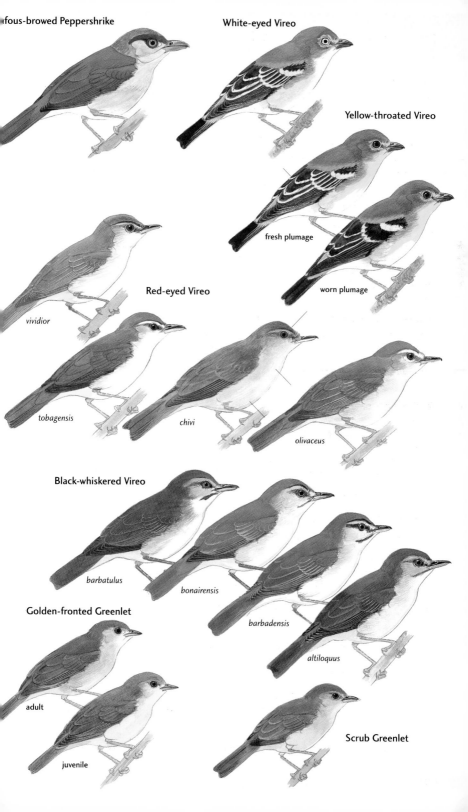

fous-browed Peppershrike

White-eyed Vireo

Yellow-throated Vireo

fresh plumage

worn plumage

*vividior*

Red-eyed Vireo

*tobagensis*

*chivi*

*olivaceus*

Black-whiskered Vireo

*barbatulus*

*bonairensis*

Golden-fronted Greenlet

*barbadensis*

*altiloquus*

adult

juvenile

Scrub Greenlet

# SWALLOWS AND MARTINS – HIRUNDINIDAE

Long wings, usually sleek bodies and sometimes extremely long tails, they are aerial feeders. Eight species treated here include both residents and migrants from North and South America.

### White-winged Swallow *Tachycineta albiventer* 14cm
Crown, back and wing-coverts glossy, greenish blue-black, white rump and tail-coverts; greater coverts and secondaries broadly fringed white; underparts white. Juvenile duller, with little gloss and little to no white on wings. **Voice** A loud *tsirrip*. **SS** From Blue-and-white Swallow by white undertail-coverts. **Status** Common resident in lowlands of C, W and S Trinidad. Frequently perches on wires. Locally uncommon in SW Tobago.

### Caribbean Martin *Progne dominicensis* 19cm
Structure as Grey-breasted Martin. Head, upperparts, breast and flanks glossy blue-black; central breast to undertail-coverts white, undertail dark. Female similar but dark grey-brown instead of blue. Juvenile like dull, drab female. **Voice** High-pitched liquid warble. **SS** Adult male obvious. Most females/immatures inseparable from Grey-breasted, but note clear cut division on underparts. **Status** Very common breeding visitor to lowland Tobago, usually present early Feb–late Oct. Rare NE and SE Trinidad.

### Grey-breasted Martin *Progne chalybea chalybea* 18cm
Chunky with triangular wings and deeply indented tail. Dark blue-black above, face and throat washed dusky, underwing-coverts the same; rest of underparts dull white, streaked on undertail-coverts. Female duller, less glossy above; juvenile sooty-brown above. **Voice** A rather rich *churr*. **SS** See Caribbean Martin. **Status** Very common widespread Trinidad resident. One old record for Tobago.

### Blue-and-white Swallow *Notiochelidon cyanoleuca cyanoleuca* 12.5cm
Small with shortish broad wings, dark metallic blue head, body-sides and undertail-coverts; underparts pure white. Juvenile brown, not blue. **Voice** A thin, high-pitched *zee*. **SS** Adult from White-winged by lack of white on rump and wings, and dark undertail-coverts. Juvenile from Southern Rough-winged by dark brown throat mottling, dark undertail-coverts and dark brown upperparts. From Bank Swallow by lack of breast-band and dark undertail-coverts. **Status** Uncommon visitor C and W Trinidad, favouring open areas, Apr–Oct. No records on Tobago.

### Southern Rough-winged Swallow *Stelgidopteryx ruficollis aequalis* 14cm
Small and dainty with peaked crown and slightly notched tail. Brown above with paler rump, throat apricot-buff, greyish-olive breast, and rest pale buff. Juvenile has pale fringes above. **Voice** A plain buzz and louder, sweet *chirrip*. **SS** Throat and rump colours diagnostic. **Status** Common, widespread Trinidad resident of open habitats and forest clearings, frequently perches on wires. Rare on Tobago.

### Cliff Swallow *Petrochelidon pyrrhonota* (subspecies unknown) 14cm
Compact with triangular wings, pale rump and square-ended tail. White forehead, blue-black crown, rufous head-sides and throat, and patch between throat and breast black; nuchal collar to upper breast pale greyish, rest of underparts creamy, greyish on undertail-coverts; blackish above with pale cinnamon rump. Juvenile lacks distinctive white forehead and throat patch, and is duller. **Voice** Low-pitched *churr*. **SS** From Barn Swallow by tail shape. **Status** Rare visitor to inland wetlands on both islands, Sep–Apr.

### Bank Swallow (Sand Martin) *Riparia riparia riparia* 12.5cm
Small-headed and slender-winged with slightly forked tail. Brown above, white below, with brown breast-band. Juvenile has pale cinnamon fringes above and buffy tinge below, both soon lost. **Voice** A dry buzzy, scratchy chatter. **SS** From juvenile Blue-and-white by brown breast-band and white undertail-coverts. From Southern Rough-winged by white throat and uniform upperparts. **Status** Scarce visitor Trinidad Sep–Apr. Rare on Tobago.

### Barn Swallow *Hirundo rustica erythrogaster* 18cm
Long-winged with long, deeply forked tail. Deep blue-black above, with rufous forehead and throat, cinnamon to buffy underparts, long, forked tail with white discs on underside. Juvenile paler, especially on face. **Voice** A dry sharp *clik*. **SS** From Cliff Swallow by tail shape and lack of rump patch. **Status** Common widespread visitor to lowlands of T&T. Some year-round, but mostly Aug–early Jun.

White-winged Swallow

Caribbean Martin

♀

♂

Grey-breasted Martin

juvenile

adult

Blue-and-white Swallow

adult

juvenile

Southern Rough-winged Swallow

Cliff Swallow

adult

adult

immature

juvenile

Bank Swallow/Sand Martin

Barn Swallow

adult

immature

## WRENS – TROGLODYTIDAE

Just two occur in T&T. Typically small with rounded wings and slightly decurved bills, whilst skulking they are very vocal, particularly inquisitive and quick to scold a potential threat.

### Rufous-breasted Wren *Thryothorus rutilus rutilus*     12.5cm

Olive-brown above with irregular barring on tail, face to neck-sides spotted, streaked and barred black and white. Breast, sides and flanks rufous, thighs olive, belly white with fine black spots, undertail-coverts barred black and olive. Juvenile paler. *T. r. tobagensis* (Tobago) brighter with paler rufous breast and buff belly. **Voice** Alarm a short dry throaty rattle. Song a clear ringing musical whistle, *hu-ee hui h-wee-uu*. **SS** None. **Status** Common, widespread Trinidad resident of cultivated estates and forest. Less common, in hillside forest, on Tobago.

### Southern House Wren *Troglodytes musculus albicans*     12.5cm

Brown above with black barring on wings and tail, pale eyebrow; whitish below, flushed buffy on flanks, barred buffy and dusky on undertail-coverts. Race *tobagensis* (Tobago) brighter, barred on rump to tail, washed rufescent on sides, to lightly barred undertail-coverts. Juveniles more richly coloured with orange-rufous wash below. **Voice** A harsh *krrrik* or typical buzzing chatter. Song comprises varied rich fruity notes, e.g. *chewee chu chewee*. **SS** None. **Status** Common widespread resident on both islands.

## GNATWRENS – POLIOPTILIDAE

Typically active and noisy, tail usually cocked and wagged.

### Long-billed Gnatwren *Ramphocaenus melanurus trinitatis*     12.5cm

Tiny, with long bill and long, erect tail. Head gingery-brown, with white chin to undertail-coverts. Greyish-olive above with darker wings and tail. Juvenile browner above. **Voice** Long, dry throaty rattle. **SS** None. **Status** Common Trinidad resident, favours dense tangles in forests at all elevations. Absent from Tobago.

## MOCKINGBIRDS – MIMIDAE

All mockingbirds share grey, brown or bluish body plumage, some with white face or wing markings.

### Tropical Mockingbird *Mimus gilvus tobagensis*     25cm

Short, often-drooped wings and long tail distinctive. Adult has buff-grey head with broad white supercilium, darker grey ear-coverts and orange-yellow eyes. Underparts grey-white. Above mid-grey, rump paler. Tail dark grey-brown, tipped off-white. Wings dark grey-brown, fringed white. Bill and legs dark grey. Young have dark eyes. **Voice** A harsh, low-pitched *chi* or *che*. Song highly variable: a mix of rich and fruity *churr*, *cherp* and whistled notes. **SS** None. **Status** Very common widespread resident on both islands. Absent from dense forest.

## WAGTAILS – MOTACILLIDAE

Slim, long-tailed ground feeders with habitual constant tail-pumping action.

### White Wagtail *Motacilla alba*     18cm

Very dainty bird that walks and runs on ground in search of insects, with jerky head movements and longish black tail constantly wagged up and down every time bird stops. Grey above, white below, with black bill and pale grey legs. Single record not identified to subspecies level; two distinct races possible. *M. a. alba*, from Europe, in winter plumage has grey head with white face and underparts, and irregular black crescent on breast. Race *ocularis*, from North America, in winter plumage has crown black, a clear black eyeline, bib-like breast crescent and solid white wing-bar. **Voice** Single disyllabic *twizzik*. **SS** None. **Status** Accidental on C Trinidad, male in Dec 1987.

**Rufous-breasted Wren**

*rutilus*

juvenile

*tobagensis*

adult

**Southern House Wren**

*tobagensis*

adult

*rutilus*

*albicans*

adult

juvenile

**Long-billed Gnatwren**

**Tropical Mockingbird**

**White Wagtail**

*alba*

*ocularis*

# THRUSHES – TURDIDAE

Some 178 species occur worldwide. Five resident species on Trinidad, three of which also occur on Tobago. In addition to the five resident species, two boreal migrant *Catharus* species have rarely been recorded. Predominantly plain-plumaged, black, brown, grey and white, some with distinctive orbital markings. Most are round-headed, long-winged and long-tailed.

## Orange-billed Nightingale-Thrush *Catharus aurantiirostris birchalli*      17cm

Slight, plump-bellied and short-tailed with horizontal gait. Adult has rufous-brown crown. Face grey with thin, bright red orbital. Chin and throat off-white, breast dark grey, paler on belly. Upperparts warm chestnut-brown, wings slightly brighter. Bill slightly upturned and more red than orange; legs more orange than red. Immature olive-brown above, with broad, teardrop-shaped orange terminal spots, from forecrown to back and on median wing-coverts; greater wing-coverts, tertials and uppertail-coverts fringed orange, underparts appear scaled from throat to flanks, with pale creamy undertail-coverts. Bill dull horn; legs yellowish-horn. **Voice** Song a repeated thin, sweet jangle of 9–10 notes. Alarm a harsh wren-like *bray* repeated 5–6 times. **SS** None. **Status** Uncommon Trinidad resident in hill forest, rarely below 650m. Keeps low to ground but sings from exposed snags. Absent on Tobago.

## Grey-cheeked Thrush *Catharus minimus minimus*      16cm

Entirely greyish-olive above, with distinctive grey-mottled ear-coverts and thin white orbital. White below, with greyish wash and small black blotches on upper breast. Bill black with yellow base, legs vinous pink. *C. m. aliciae* slightly larger and washed brownish-olive on flanks. **Voice** Usually silent. **SS** From Veery by colder olive-grey head and upperparts, and darker streaking on the breast. **Status** Accidental on Trinidad with just four records, Nov–Mar, and none since 1989. No sightings on Tobago.

## Veery *Catharus fuscescens fuscescens*      17cm

Adult has warm chestnut-brown head with greyish ear-coverts. Throat and upper breast buff-white with warm brown mottling. Rest of underparts grey-white with diffuse flank mottling. Upperparts chestnut-brown with darker flight feathers and tail. Bill black, legs grey-pink. *C. f. salicicola* slightly more olive above, pale grey below, with arrowhead mesial and breast streaks, washed grey, and diffuse flanks streaks; rest of underparts white. **Voice** Usually silent. **SS** From other *Catharus* by brown mottling restricted to upper breast, whiter lower underparts and warmer upperparts. **Status** Accidental on Trinidad, with just two records, both from Northern Range, Apr 1975 and Oct 1982. No sightings on Tobago.

## Yellow-legged Thrush *Turdus flavipes*      22cm

Two distinct races, one on each island. Adult male *P. f. melanopleura* (Trinidad) has black head with yellow orbital. Throat and upper breast black, dark grey lower breast and belly, and paler undertail-coverts. Upperparts smoky grey, wings and tail black. Bill and legs yellow. Adult male *T. f. xanthoscela* (Tobago) all black with yellow orbital, bill and legs. Females of both races have olive-brown head with faint white orbital. Breast and belly paler olive with yellow flanks. Upperparts dark olive-brown. Bill dirty yellow in some, browner or almost black in others. Legs pale straw-yellow. Immature recalls respective adult, but with mottled underparts and buff tips to mantle and wing-coverts. **Voice** A thin high-pitched *seep*. Song very varied and musical. **SS** Male unmistakable. Female from smaller White-necked Thrush by lack of throat streaking and white collar, and darker underparts. **Status** Uncommon resident of high-altitude forest on both islands.

Veery

*fuscescens*

Grey-cheeked Thrush

Orange-billed Nightingale-Thrush

juvenile

adult

Yellow-legged Thrush

juvenile

♀

♂

♂

*xanthoscela*
(Tobago)

*melanopleura*
(Trinidad)

## Bare-eyed Thrush *Turdus nudigenis nudigenis* 24cm

Adult has olive-brown head with broad orange-yellow orbital. Chin and throat off-white variably streaked black. Breast and upper belly pale brown with grey-white lower belly and undertail-coverts. Nape and mantle olive-brown, wings and tail dark grey. Bill dull yellow with creamy white tip; legs yellowish-pink. Juvenile much more rufescent than adult, and lacks any greyish washes, with orange shaft-streaks above, and terminal spots on wing-coverts and tertials. **Voice** A distinctive, feline *keea eee* or *rur-rii*, second syllable higher. Song a series of repetitive notes, like Cocoa Thrush but less musical, more melancholy. **SS** None. **Status** Very common, widespread resident on both islands.

## White-necked Thrush *Turdus albicollis phaeopygoides* 20cm

Adult has dark greyish-brown head and upperparts, becoming greyish on uppertail-coverts and tail. Chin and throat black with white streaks, upper breast pure white, breast to flanks and belly soft grey, thighs barred brown and grey, and undertail-coverts white. Bill dark above, yellowish below; legs greenish-yellow. Juvenile has orange shaft-streaks from crown to lower back and terminal spots on wing-coverts. Full white chin and throat is slightly spotted brown, whilst upper breast and sides have patchy orange and (smaller) black spots, fading on grey lower breast to flanks. **Voice** Song a series of slow, haunting double or paired notes; alarm a low-pitched sharp *chup*. **SS** None. **Status** Common resident in forest on both islands, but rather shy. Tends to feed close to the ground and often hops along forest trails in early morning.

## Cocoa Thrush *Turdus fumigatus aquilonalis* 23cm

Large, heavily built thrush, warm rufous-brown above, greyish on head-sides, grey orbital ring; cinnamon below, paler on throat and heavily scalloped white on undertail-coverts. Bill dark, legs flesh-pink. Juvenile has grey orbital ring but lacks grey wash on head-sides; dark fringes to all feathers from crown to back give scaled appearance, scapulars have pale shaft-streaks, and wing-coverts orange terminal spots. Darker below, especially undertail-coverts. **Voice** Song musical and variable, with loud, highly repetitive series of short couplets, *cheer-hoo cheer-hoo*. A harsh scolding *chat a chat chat*. **SS** None. **Status** Common Trinidad resident, in gardens, estates and forest. Absent on Tobago.

Bar

White-necked Thrush

juvenile

adult

adult

juvenile

Cocoa Thrush

adult

juvenile

adult

## NAGERS – THRAUPIDAE

A diverse family of 231 species found throughout the New World. Several are long-distance migrants; the majority resident. Conebills are small, warbler-like, drab and dull in appearance. Just one species occurs on Trinidad. Honeycreepers are relatively slim, often with decurved bills. Five species occur on Trinidad with one also on Tobago. Tanagers are more rounded with larger bills and 14 species have been recorded on Trinidad, two of which are migrants, yet only five occur on Tobago.

### White-shouldered Tanager *Tachyphonus luctuosus flaviventris* 14cm

Adult male black with conspicuous white shoulder-patch and white underwing-coverts. Bill black, legs grey. Female has grey head. Throat off-white, rest of underparts dirty lemon-yellow. Nape grey, rest of upperparts olive-green, with darker tail. Bill grey. **Voice** A quiet *chut* or *chup*; song a high-pitched thin *sweet sweet sweet*. **SS** See White-lined Tanager. From White-winged Becard by smaller size, more delicate jizz, solid white wing-patch and lack of white undertail chevrons and tip to tail. **Status** Uncommon Trinidad resident of low- and mid-elevation forests. Absent from Tobago.

### White-lined Tanager *Tachyphonus rufus* 19cm

Adult male all black, with white axillaries normally only noticeable when wings raised or in flight. Upper mandible lead grey, lower silver-grey. Legs grey. Female rich chestnut-brown, slightly darker on crown and flight feathers. Juvenile like paler female, light cinnamon on breast; immature males progressively acquire black. **Voice** A resonant *chip*, metallic *chink* and rather drawn-out, sweet *tsip*. Song a quick series of disyllabic chirps. **SS** Male from White-shouldered Tanager by larger size and white restricted to underwing linings. From Shiny Cowbird by rounder crown and black not iridescent plumage. Female from female Crowned Ant Tanager by chestnut not olive-brown plumage, and no crown patch. **Status** Very common resident widespread on Trinidad, but less numerous on Tobago.

### Silver-beaked Tanager *Ramphocelus carbo magnirostris* 18cm

Adult male has deep red 'crushed velvet' head and upper breast; lower breast and belly slightly darker. Upperparts duller and darker, wings and tail dark grey-brown. Bill has unique blue-white plate covering lower mandible, upper is dark grey. Legs dark. Female has reddish-brown underparts, darker brown head, mantle, wings and tail with contrasting reddish-brown rump. Bill dark lead grey. **Voice** A sharp *chip*, song a rich series of 2–3-syllable notes, *chik wich* or *chu chu tweep*. **SS** None. **Status** Common widespread Trinidad resident, especially fond of bamboo. Absent on Tobago.

### Blue-grey Tanager *Thraupis episcopus nesophilus* 18cm

Adult male has grey-white head and throat with bare-faced expression and beady black eyes. Underparts blue-grey. Nape and mantle mid-grey, rump brighter blue. Tail slightly notched, bright turquoise-blue except black central feathers. Lesser coverts lilac, rest of coverts and flight feathers bright turquoise-blue. Tertials black fringed blue. Bill and legs grey. Female slightly duller. Juvenile duller and greyer with few blue tones. *T. e. berlepschi* (Tobago) brighter and richer blue on wing-coverts and rump. **Voice** Erratic series of high-pitched squeaks. Most common call a high-pitched, drawn-out *siiur*. **SS** None. **Status** Very common and widespread resident in T&T.

### Palm Tanager *Thraupis palmarum melanoptera* 18cm

Adult in fresh plumage has yellow-olive crown, olive-grey nape and face. Breast olive-grey becoming brownish on belly and undertail-coverts. Mantle and rump slightly darker olive-grey with paler and greener wing-coverts and shoulders. Flight and tail feathers dark grey-brown with pale grey-green wing-stripe. Bill and legs grey. Juvenile recalls dull adult. **Voice** Song a high-pitched series of sweet yet scratchy, squeaky notes, e.g. *weet a weet a weet weet weet* or *huet a wert a tzee tzee tzee*; flight-call a high-pitched *sweeet*. **SS** Unlikely to be confused. **Status** Abundant and widespread resident on Trinidad; less numerous but still common on Tobago.

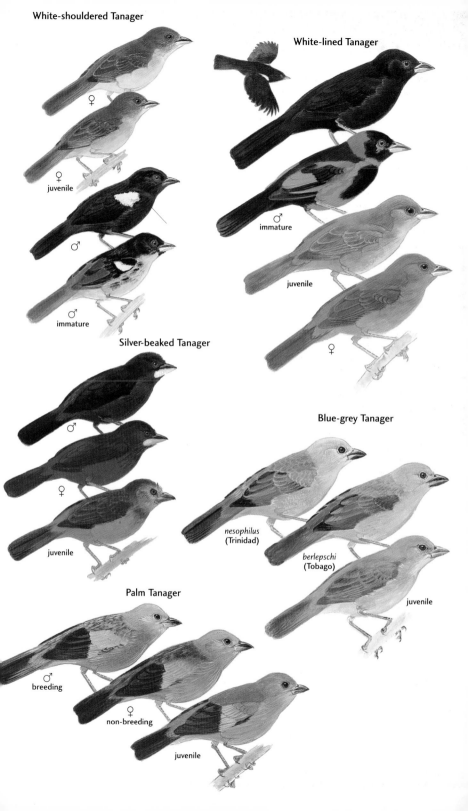

**White-shouldered Tanager**

♀

♀ juvenile

♂

♂ immature

**White-lined Tanager**

♂ immature

juvenile

♀

**Silver-beaked Tanager**

♂

♀

juvenile

**Blue-grey Tanager**

*nesophilus* (Trinidad)

*berlepschi* (Tobago)

juvenile

**Palm Tanager**

♂ breeding

♀ non-breeding

juvenile

## Blue-capped Tanager *Thraupis cyanocephala buesingi* 19cm

Adult has bright dark blue head with black lores, black immediately above bill and pale grey moustachial. Throat, breast and upper belly olive-grey, paler on lower belly with dull yellow thighs. Undertail-coverts yellow. Upperparts bright olive-green with yellow underwing-coverts and patch at wing bend. Bill black, legs blue-grey. Juvenile is duller. **Voice** Usually quiet, song is a subdued series of squeaks. **SS** None. **Status** Scarce resident of forest, principally above 650m, on Trinidad; absent from Tobago.

## Turquoise Tanager *Tangara mexicana vieillioti* 14cm

Adult has deep bright blue forecrown, and face, throat and upper breast blotched black. Lower breast, flanks and undertail-coverts pale lemon-yellow with heavy black and blue flanks blotching. Rear crown, nape and fore mantle black, lower mantle and rump bright blue. Outer tail feathers blue, rest black. Shoulders bright blue with remaining covert and flight feathers black. Bill and legs black. Juvenile has unmarked dirty cream underparts with dull purple breast wash and is browner above. **Voice** Very soft, high-pitched twittering. **SS** None. **Status** Common Trinidad resident of forest and cultivated estates. Absent from Tobago.

## Bay-headed Tanager *Tangara gyrola viridissima* 14cm

Adult has plum red head; underparts bluish-green. Upperparts apple green with golden nuchal collar and brighter lime-green wing-coverts. Bill darker horn above, paler below. Legs vinous. Immature has dirty yellow-brown head, paler green underparts and duller upperparts. **Voice** A fairly high-pitched scratchy *siiawiii*, a buzzy chatter and soft *spss*. **SS** None. **Status** Very common Trinidad forest resident. Absent from Tobago.

## Speckled Tanager *Tangara guttata trinitatis* 14cm

Adult has apple green crown spotted black, with yellow forecrown and face, and black lores. Throat, breast and belly grey-white heavily spotted black, with rear flanks tinged green. Nape, mantle and uppertail-coverts apple green with black speckling; rump unmarked green. Outertail-feathers green, rest dark grey. Wing-coverts and tertials black fringed peppermint-green; flight feathers grey with aquamarine fringes to secondaries. Bill silver-grey with black cutting edges. Legs blue-grey. Immature duller with smudging rather than speckling. **Voice** Quiet high-pitched *ssst*. **SS** None. **Status** Uncommon resident of high forest edge in Trinidad, rarely seen below 600m. Absent from Tobago.

## Swallow Tanager *Tersina viridis occidentalis* 15cm

Adult male has black forehead, face and throat. Breast and flanks bright turquoise-blue with bold black chevrons on rear flanks. Central belly and undertail-coverts white, undertail dark grey. Upperparts turquoise-blue, flight feathers and tail black with extensive blue fringes. Bill and legs black. Female has olive-grey face and throat. Upper breast green becoming greenish-yellow on central breast and belly. Flanks dull lemon-yellow with olive chevrons. Crown and upperparts apple green with duller tail; flight feathers black broadly fringed green. Immature male initially green mottled variably blue. Takes several years to attain full adult plumage. **Voice** A sharp disyllabic *zü eer*, and series of single squeaky notes. **SS** None. **Status** Scarce breeding visitor to Trinidad, found mid-Feb–late Aug in Northern Range, usually above 600m. Absent from Tobago.

## Blue Dacnis *Dacnis cayana cayana* 12.5cm

Rather small and short-billed. Adult male has bright blue head with black eyeline and red eyes. Throat black, rest of underparts blue. Nape, mantle and tail black, rump bright blue. Wings black fringed bright blue. Bill, thin and steel grey. Legs pink. Female has bright blue head, grey-white throat and apple green underparts. Upperparts green, tail black. Wings greenish-olive, fringed brighter green. Immature similar to female with variable black mask and bib. **Voice** High-pitched *tsit*. **SS** Male unmistakable. Female from Green Honeycreeper by smaller size, blue head and all-dark, thin bill. **Status** Uncommon Trinidad resident in cultivated estates, second growth and forest edge. Absent from Tobago.

## Blue-capped Tanager

adult

juvenile

## Turquoise Tanager

adult

juvenile

## Bay-headed Tanager

adult

juvenile

## Speckled Tanager

adult

juvenile

## Swallow Tanager

♂

immature ♂

♀

## Blue Dacnis

♂

immature ♂

♀

### Red-legged Honeycreeper *Cyanerpes cyaneus cyaneus* 12.5cm

Dumpy with slightly decurved bill. Adult male has pale turquoise-blue crown, black lores and mask. Throat, breast and belly deep blue. Wraparound shawl and fore mantle black; rest of upperparts deep blue with black distal tail feathers. Wing-coverts black with broad blue scapular line. Underwing greenish-yellow. Bill black, legs bright red. Female has grey-green head with dirty white supercilium. Breast and belly dull green streaked darker. Mantle and rump grey-green, wings and tail darker. Legs reddish-brown. Male in non-breeding plumage like female, but has black wings and tail. Juvenile male changes from looking like female to non-breeding male, but acquires blue patches very quickly. *C. c. tobagensis* (Tobago) has straighter, heavier bill. **Voice** High-pitched, thin *tsip* and short, rather harsh *shiir*. **SS** See Purple Honeycreeper. **Status** Common resident in T&T, in forest and wooded estates.

### Purple Honeycreeper *Cyanerpes caeruleus longirostris* 11.5cm

Dumpy, short-winged with long, decurved bill. Adult male has deep purple-blue body with black forehead, lores, mask, throat, wings and distal half of tail. Bill black, legs bright yellow. Female has dull green crown. Face and throat dirty pink, ear-coverts narrowly streaked black with blue malar. Breast and belly greenish-white with dark green streaks and buff-pink wash to upper breast. Upperparts green, wings and tail greyer. Legs dull greenish-yellow. Juvenile recalls adult female with fewer underparts streaks. **Voice** Varied thin high-pitched notes, most regularly *dswee*. **SS** Male from Red-legged by deep purple crown, throat, underwing and leg colour. Female by face, underparts and leg colour. **Status** Common Trinidad resident of forest and wooded estates. No recent records on Tobago.

### Bicoloured Conebill *Conirostrum bicolor bicolor* 11.5cm

Small and nondescript. Adult male has blue-grey head with dull red eyes. Underparts dull fawn-grey. Upperparts blue-grey, wings browner. Upper mandible dark grey, lower pinkish. Legs orange-pink. Female duller. Immature has olive-grey head; underparts greenish-yellow with grey wash on flanks. Upperparts olive-grey. Adult plumage not acquired until second or third calendar year, but often breeds in immature plumage. **Voice** High-pitched, thin twittering *tsee tsee*. **SS** Adult unlikely to be confused. Juvenile from similar female Yellow Warbler by stubbier jizz, dirtier underparts, lack of bright flight-feather fringes and dull concolorous tertials. **Status** Common resident in Trinidad's mangroves. Absent Tobago.

### Green Honeycreeper *Chlorophanes spiza spiza* 14cm

Large and comparatively stout-billed. Adult male has complete black hood with red eyes. Throat and upper breast bright green. Lower breast and belly jade-green becoming grey-white on undertail-coverts. Upperparts jade-green, wings and tail darker green with black centres to tertials and tail feathers. Bill slightly decurved; black upper mandible and bright yellow lower. Legs grey. Female has apple green body with yellower throat, lower belly and undertail-coverts. Central tail black, rest green. Bill more extensively yellow. Juvenile duller with shorter, dark grey bill and legs, and brown eyes. **Voice** Scolding *chip chip* and loud *tswee tswee*. **SS** Male unmistakable. Female from female Blue Dacnis by larger size, green head and yellow bill. **Status** Common Trinidad resident, in woodland, estates and second growth at all elevations. Absent Tobago.

## GENUS INCERTAE SEDIS

The taxonomic affinities of *Coereba* are uncertain. It may be closely related to *Tiaris* or the sole representative of its own family.

### Bananaquit *Coereba flaveola luteola* 10cm

Tiny yet large-headed. Adult has black crown with white supercilium from bill to nape. Lores, eyeline and upper ear-coverts black, lower ear-coverts, chin and throat pale grey. Breast and belly lemon-yellow grading off-white on flanks and undertail-coverts. Nape and fore mantle black, lower back and rump bright lemon-yellow. Tail black. Wings black with conspicuous but small white patch at base of primaries. Bill black and slightly decurved, legs grey. Immature duller with yellow supercilium and grey-washed underparts. **Voice** Song a high-pitched twittering, including a repeated *tzee tzee tzee*. A sharp *tseep*. **SS** None. **Status** Possibly the most widespread and abundant passerine in T&T.

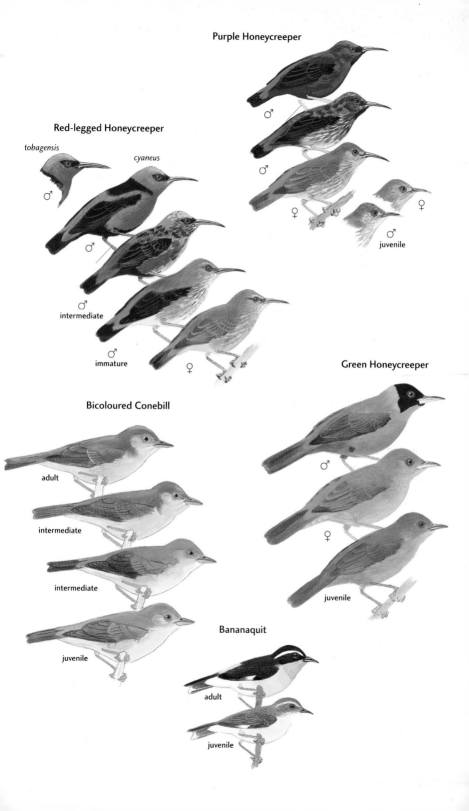

**Purple Honeycreeper**

♂

♂

♀

♀

♂
juvenile

**Red-legged Honeycreeper**

*tobagensis*

*cyaneus*

♂

♂

♂
intermediate

♂
immature

♀

**Green Honeycreeper**

**Bicoloured Conebill**

adult

intermediate

intermediate

♂

♀

juvenile

juvenile

**Bananaquit**

adult

juvenile

## GENUS INCERTAE SEDIS

The placement of *Tiaris* in Emberizidae is no longer taken for granted and the genus is currently pending further genetic evaluation.

### Black-faced Grassquit *Tiaris bicolor omissa*      11.5cm

Old male has sooty-black head and underparts, with undertail-coverts fringed buffy; upperparts olive-green. Female soft olive-green above, paler, buffy-green below. Juvenile like female. **Voice** Thin squeaky buzz. **SS** Male unmistakable. Female from Blue-black Grassquit by unstreaked underparts. **Status** Very common, widespread lowland resident on Tobago. Only on Bocas Is, Trinidad.

### Sooty Grassquit *Tiaris fuliginosus fumosus*      10cm

Old male is entirely sooty-black with slightly glossy, olive-green sheen. Female dusky olive-brown and quite undistinguished; dark horn bill. Juvenile like female but much greener, with distinct two-tone bill. **Voice** Thin high-pitched jangle. **SS** See Blue-black Grassquit. **Status** Uncommon Trinidad resident, in forest edge usually above 400m. Absent from Tobago.

## SEEDEATERS AND ALLIES – EMBERIZIDAE

Seedeaters have all but disappeared from T&T due to the cagebird trade. Just two are regular on Trinidad; others very rarely occur on either island. Four species of finches occur on Trinidad. Of the three yellow-plumaged birds one is resident, one has very recently colonised and one is a vagrant.

### Blue-black Grassquit *Volatinia jacarina splendens*      10cm

Adult male almost entirely glossy blue-black, with small white patch at inner bend of wing. In fresh plumage shows variable buffy fringes. Female soft brown above, buffy-white below with soft streaking on breast. Juvenile like female, but warmer below and less heavily streaked on breast. **Voice** A single short *psiieu* whilst springing upwards before landing on same perch. **SS** Male from Sooty Grassquit by glossy plumage and white axillaries. Female from Sooty and Black-faced Grassquits by breast streaking. **Status** Abundant, widespread Trinidad resident; common in lowland Tobago.

### Grassland Yellow Finch *Sicalis luteola luteola*      11.5cm

Slender, graceful and shy, adult male is olive-green above with blackish streaks, particularly dark on wings and tail; yellow face and entire underparts. Female has no yellow on face, has whitish throat, and rump/uppertail-coverts more greenish. Juvenile slightly duller than female and streaked on breast. **Voice** Trilling buzz. Song a series of elongated *chuuze* notes uttered both from perches and in parachuting flight. **SS** None. **Status** Flock of up to 50 present in C Trinidad since Mar 2004, where favours open, grassy fields. Absent from Tobago.

### Orange-fronted Yellow Finch *Sicalis columbiana colombiana*      12.5cm

Adult male yellowish olive-green above, subtly streaked, with some yellowish fringes, and orange forehead; yellow below with slight greenish wash to breast. Female yellowish-buffy above, well streaked dusky; pale buffy below with soft streaking. Juvenile is slightly paler version of female, with a pale bill. **Voice** Series of short *chip* notes. **SS** See Saffron Finch. **Status** Accidental on Trinidad with just one sighting, of several birds, in 1926.

### Saffron Finch *Sicalis flaveola flaveola*      14cm

Adult male all yellow, washed orange over crown; back to uppertail-coverts washed green, flight and tail feathers blackish, edged yellow. Face to undertail-coverts yellow. Bill black with pale blue base. Female similar but streaked above. Young buffy-grey with dusky streaks on back, wings and tail dusky with yellowish-buffy fringes, pale to whitish below, flushed yellow on breast with some dusky streaking. **Voice** Dry chattering *tzip* and *dzeep* notes. A sweet, loud *sirrup sirrup sirrup*. **SS** Male from smaller Orange-fronted Yellow Finch by brighter yellow underparts. **Status** Locally common resident in W Trinidad restricted to areas of short grass. Introduced, but not seen for many years on Tobago.

**Blue-black Grassquit**

**Black-faced Grassquit**

old male

♂ breeding

♂ after postnuptial moult

♀

♂ juvenile

♂

♂

♀

**Sooty Grassquit**

♀

♂

juvenile

**Grassland Yellow Finch**

juvenile

♂

♀

**Orange-fronted Yellow Finch**

♂

♀

juvenile

**Saffron Finch**

♂

♀

♂ immature

♀ immature

## Slate-coloured Seedeater *Sporophila schistacea longipennis*    11.5cm
Adult male has dark grey head, throat and upper breast with variable yellow or white orbital and white malar. Throat dark grey, rest of underparts white with greyer flanks. Upperparts dark grey, almost black on wings and tail, with white speculum and very faint white wing-bar. Bill orange, legs olive. Polymorphic, with many intermediates between the morphs shown on the plate. Female has dark olive-brown head. Upper breast buff-brown, lower breast and belly buff-white. Upperparts dark grey, legs olive. Immature recalls adult female with yellowish belly; male may have orange bill. **Voice** Song a sweet but rather plain trill, lacking musical variation of Grey Seedeater. **SS** Grey morph adult male separable with extreme care from Grey Seedeater by darker body, malar stripe (if apparent), speculum, wing-bar and less bulbous bill. Female and immature inseparable in the field. **Status** Once an uncommon resident on Trinidad, now locally extirpated with no reliable reports of wild birds for many years.

## Grey Seedeater *Sporophila intermedia intermedia*    11.5cm
Adult male has grey head, throat, neck-sides, upper breast and flanks. Lower breast and belly white merging in an inverted V into grey. Upperparts grey with small white speculum. Bill bulbous and yellowish-pink, legs grey. Female has dark brown crown. Face throat and upper breast rich buff-brown, lower breast and belly buff-white. Upperparts dark olive-brown. Bill dark horn, legs brownish-grey. Immature recalls adult female but may have pale bill. **Voice** Varied rich twitters, trills and chirps. **SS** Adult male separable with care from Ring-necked Seedeater by entirely greyish head and throat, and lack of whitish band on rump. Female and immature inseparable in the field. See also Slate-coloured Seedeater. **Status** Formerly very common resident in open grassy areas on Trinidad, now locally extirpated with no documented sightings of wild birds for at least ten years.

## Wing-barred Seedeater *Sporophila americana americana*    11.5cm
Adult male has black head with white lower eye crescents. Throat white extending as neck-sides collar. Rest of underparts grey-white with diffuse grey breast-band. Upperparts black with two white wing-bars, tertial tips and speculum, and pale grey rump. Bill stout and black, legs black. Female has dark olive-brown head. Throat dirty-white, breast buff-brown, paler on belly. Upperparts dark olive-brown. Bill dark horn, legs grey. Immature recalls adult female; male has darker wings. **Voice** Series of trills and whistles. **SS** Adult male from Lesson's Seedeater by larger bill, white throat, breast-band, wing-bars and rump colour. Immature and female separable with extreme care from Black-faced Grassquit and other seedeaters by very dark brown upperparts and buffier underparts. **Status** Former Tobago resident, now probably locally extirpated with no documented records for many years.

## Ring-necked Seedeater *Sporophila insularis*    11.5cm
Adult male has grey head, throat, upper breast and flanks with variable whitish malar and band across throat. Rest of underparts white merging in an inverted U into grey. Upperparts mid-grey with white tips to innermost median coverts, white speculum and whitish band on lower rump, that only shows when bird is at rest or singing. Bill bulbous and yellowish-pink, legs grey. Female and immature identical to Grey Seedeater. **Voice** Varied twitters, trills and chirps, less complex than Grey Seedeater. **SS** See Grey Seedeater. **Status** Formerly a common resident of light woodland on Trinidad, now locally extirpated with no documented sightings of wild birds for many years.

Slate-coloured Seedeater

♂ adult
olive-citron morph

♂ immature

♀ (and juvenile)

normal (grey) morph

♂

♂ old adult

normal (grey) morph

Grey Seedeater

♂ adult

♂ first year

♂ immature

♀

juvenile

Wing-barred Seedeater

♂

♀

Ring-necked Seedeater

♂

♂ 1st basic

## Lesson's Seedeater *Sporophila bouvronides* 11.5cm

Adult male has black head with broad white malar. Throat black, more mottled on upper breast. Rest of underparts white with faint dark flanks streaking. Upperparts black with white speculum and rump. Bill and legs black. Female has olive-brown head. Throat and upper breast buff-brown, paler on lower breast and belly. Upperparts olive-brown. Upper mandible black, lower pale horn. Immature male whiter below and more greyish above than female; juvenile like female. **Voice** Long trill and various chatter notes. **SS** From Wing-barred Seedeater by slighter build and smaller bill. Male by white malar, black throat, white speculum and lack of breast-band and wing-bars. Immature and female inseparable from other *Sporophila*. **Status** Former resident now locally extirpated. Rare visitor to lowland scrub on W Trinidad and Bocas Is in Aug–Oct. No recent sightings on Tobago.

## Lined Seedeater *Sporophila lineola* 11.5cm

Adult male very similar to male Lesson's, but has bold white crown-stripe, black chin and unmarked white underparts. Female and immature inseparable from Lesson's. **Voice** Birds reaching Venezuela (and possibly T&T) from south do not sing. Those reaching Guyana and Delta Amacuro of Venezuela (and possibly T&T) from east do sing. **SS** See Lesson's, Yellow-bellied, and Ruddy-breasted Seedeaters. **Status** Accidental. A flock of 33 birds found in SW Trinidad, Sep 2007.

## Yellow-bellied Seedeater *Sporophila nigricollis nigricollis* 11.5cm

Adult male has black head, throat and upper breast. Rest of underparts from pale yellow to almost white, sometimes with faint flanks streaking. Upperparts olive with browner wings and tail; occasionally has small white speculum at base of primaries. Bill and legs pale grey. Female has olive-brown head and upperparts. Underparts buff-yellow. Bill black. Immature recalls adult female but male may show emergent black on head. **Voice** A single *cherp*; song a series of whistles, *chu chu chi pree zee zee*. **SS** Adult male unmistakable. Female and juvenile impossible to separate from other small *Sporophila*. **Status** Former resident in T&T. Now a rare visitor from mainland, mostly to NW peninsula of Trinidad and Bocas Is, May–Oct.

## Ruddy-breasted Seedeater *Sporophila minuta minuta* 10cm

Old male has grey head. Underparts variably reddish-brown, richer colour develops with age. Nape and mantle grey or brownish, rump reddish-brown. Wings and tail dusky with brown feather fringes and small white speculum. Bill and legs grey or black. Female has olive-brown head and upperparts. Throat and breast buff-brown, paler buff-white on belly. Bill horn. Immature recalls adult female with two-toned bill, paler below. **Voice** Series of quick disyllabic musical whistles, *tzu tzu, pee-wee*. **SS** Adult male unmistakable. Female and juvenile impossible to separate from other *Sporophila*. **Status** Rare and local resident of lowland savanna and freshwater marsh edge on Trinidad. Now probably extirpated on Tobago.

Lesson's Seedeater

Lined Seedeater

♂
definitive

♂
1st year

♀

♂

♀

Yellow-bellied Seedeater

♂

♂
immature

♀

juvenile

Ruddy-breasted Seedeater

♂
ɔwn type

♂
grey type

♂
1st year

♂
immature

♀

juvenile

## Chestnut-bellied Seed Finch *Oryzoborus angolensis crassirostris*   12.5cm

Large; habitually flicks wings. Adult male has black head, throat and upper breast, with rest of underparts deep dull reddish-brown. Upperparts black with small white speculum. Underwing linings white. Variable amount of white at base of inner webs of flight feathers, which may show as large patch in flight. Bill broad-based, bulbous and black. Legs black. Female has rufous-brown head and upperparts. Underparts buff-brown, underwing-coverts white. Juvenile recalls adult female with horn-coloured bill. **Voice** Varied fluty whistles and faster chattering. **SS** Adult male unlikely to be confused. Female and immature from seedeaters by larger size and much heavier bill, from Large-billed by smaller bill and white underwing linings. **Status** Former scarce resident on Trinidad, almost certainly locally extirpated. The most common cagebird in T&T, therefore, whilst species may occasionally be recorded in a wild state, sightings are subject to escapee caveat. No substantiated records from Tobago for many years.

## Large-billed Seed Finch *Oryzoborus crassirostris crassirostris*   15cm

Adult male all black with prominent white wing-patch. Bill bone-white, extremely broad-based and bulky; legs black. Female dark reddish-brown, slightly paler below without white wing markings. Bill dark grey, legs black. Juvenile recalls adult female with paler bill. **Voice** Varied and very musical whistles, given by both sexes. **SS** None. **Status** Former rare resident on Trinidad, now locally extirpated with a few undocumented reports from SE Trinidad in May–Jun. This is a common cagebird, and the escape potential is clear.

# CARDINALS, GROSBEAKS, SALTATORS AND ALLIES – CARDINALIDAE

A family of 43 species found throughout the New World. Saltators and grosbeaks are bulky with extremely large bills. Two saltators are resident on Trinidad and one grosbeak is a rare passage migrant. Buntings are somewhat smaller with short, conical bills. Just two migrants, one a vagrant to Trinidad, the other a migrant to both islands.

## Masked Cardinal *Paroaria nigrigenis*   18cm

Adult has crimson-red, slightly crested head with black mask and red eyes. Chin and throat red and form triangular patch coming to point on upper breast. Neck-sides and rest of underparts white. Upperparts black. Upper mandible black, lower white. Legs grey. Immature more untidy, with tawny-brown head and less distinct mask. Underparts grey-white, upperparts sooty-black. **Voice** Hard *chip* or *chep*. **SS** Unmistakable. **Status** Uncommon Trinidad resident, mainly in W coast mangrove and associated marshes and lagoons. Absent on Tobago.

## Indigo Bunting *Passerina cyanea*   12.5cm

Adult breeding male is ultramarine blue. Wings and tail black with broad blue fringes. Bill steel grey, short and stout. Legs black. Non-breeder has brown head with paler buff underparts showing variable amounts of blue. Nape and mantle brown, rump mostly blue. Wing-coverts brown fringed black, flight and tail feathers black fringed blue. Female like first-winter male but less blue on wings. Young have a little streaking on breast and sides. **Voice** A dry, sharp *spik*. **SS** Adult male unmistakable. Immature and female from slightly smaller seedeaters and grassquits by faint wing-bars, lightly streaked underparts and variable blue in rump and tail. **Status** Accidental Trinidad: Northern Range, Mar 1977.

## Dickcissel *Spiza americana*   15cm

Adult male has grey head with long yellow supercilium and moustachial; chin to neck-sides white with variable black patch on lower throat, breast to belly yellow, yellowish-grey sides and thighs, rest white. Brown above with black streaks and maroon smaller wing-coverts. Female duller, with brown head and white throat, yellow breast and slight streaking below. Immature paler and less heavily streaked. **Voice** Dry wheezy chatter. In flight, a single rasping *zzrt*. Communal daytime roosts located by loud endless twittering. **SS** None. **Status** Irruptive visitor to Trinidad, with most in Jan–Apr. Nocturnal roosts en masse in sugarcane, feeding in ricefields. In some years flocks numbering tens of thousands are present, in others species almost absent. Rare visitor to Tobago.

**Chestnut-bellied Seed Finch**

♂

♀

**Large-billed Seed Finch**

♂

♀

♂
juvenile

**Masked Cardinal**

immature

adult

juvenile

**Dickcissel**

♀

♂

♂
immature

**Indigo Bunting**

♂
winter

♂
first-winter

# PLATE 97: GROSBEAK AND SALTATORS

### Rose-breasted Grosbeak *Pheucticus ludovicianus ludovicianus* 19cm

Adult breeding male has black head and throat. Breast and mid-belly rose-pink, rest of underparts white. Nape and mantle black, rump white. Tail black with white spots on outer tail and uppertail-coverts. Wings black with broad white base to primaries and two white wing-bars. Pink underwing linings and white undertail spots. Bill ivory, legs dark grey. Non-breeder has brown fringes to upperparts and black flank chevrons. Female has dark brown crown with thin cream central stripe. Face grey-brown with broad creamy white supercilium, narrow black eyeline and moustachial which meet on ear-coverts. Throat buff-grey with broad white malar stripes. Rest of underparts buff-white streaked black. Upperparts grey-brown streaked black. Wings and tail black with two white wing-bars. Underwing linings yellow. Juvenile like female but immature male more boldly marked and shows emergent rosy on breast; sexed by underwing lining colour. **Voice** A soft, wheezy *wheek* and sharp metallic *pik*. **SS** None. **Status** Rare (passage?) migrant to T&T, mostly Mar–Apr and Nov–Dec.

### Greyish Saltator *Saltator coerulescens brewsteri* 23cm

Robust and chunky. Adult has dark grey head with short white eyebrow, black lores and white lower eye crescent. Broad black moustachial borders creamy white throat. Breast and belly olive-white, buff on undertail-coverts. Upperparts dark grey with blacker median coverts; flight feathers and tail darker. Bill very stocky and black, legs grey. Juvenile has olive-grey head with short lemon-yellow supercilium and eye crescent. Throat and upper breast olive-yellow, diffusely blotched darker, rest of underparts grey-white. Upperparts olive-grey with brighter green flight-feather fringes and darker tail. Bill two-toned; upper mandible dark horn, lower with varying amounts of yellow. **Voice** Call *whit chu wee* with third syllable higher pitched, or a repeated *whit chu whit chu*. **SS** Adult unmistakable. Juvenile from Streaked Saltator by larger size, more rounded crown, larger bill and fewer breast streaks. **Status** Common and widespread Trinidad resident. Absent Tobago.

### Streaked Saltator *Saltator striatipectus perstriatus* 19cm

Sleek with conical head. Adult has dirty grey head with thin off-white short eyebrow and orbital ring. Some show indistinct orange gape mark. Chin and throat white with dusky sides, rest of underparts white, variably streaked dark grey. Upperparts olive-grey with lime-green wing fringes and greyer rump and tail. Bill conical, black variably tipped orange or yellow. Legs black. **Voice** A short sharp *quit*; song a fluty *chut weeurr*. **SS** See Greyish Saltator; note jizz, head and bill shape, lack of pronounced moustachial and heavier underparts streaking. **Status** Very local resident in extreme NW Trinidad, in arid second growth and deciduous forest. Absent from Tobago.

Rose-breasted Grosbeak

♂
immature

Greyish Saltator

♀

♂

adult

juvenile

Streaked Saltator

juvenile          adult

## GENERA INCERTAE SEDIS

*Habia* and *Piranga* 'tanagers' are currently considered as of uncertain taxonomic affinities. It is fairly well accepted that *Piranga* should belong to Cardinalidae.

### Crowned Ant Tanager *Habia rubica rubra*                                   18cm

Large and long-tailed with rounded crown. Adult male has dull red head with greyish wash and obscure, bright red central crown-stripe bordered black. Throat bright rose-red, breast and belly dull greyish-red. Upperparts greyish-red, tail brighter. Bill dark grey, legs pale pinkish-brown. Female has dull olive-grey head with thin buff central crown-stripe. Breast and belly dull olive-yellow with warmer buff undertail-coverts. Upperparts dull olive-green, tail more rufous. Immature recalls adult female but darker. **Voice** A dry agitated *crix crix crix* chatter; a hoarse *srieu* regularly repeated. **SS** Male from female Silver-beaked Tanager by brighter red plumage, especially crown-stripe, throat and tail. Female from female White-lined Tanager by olive-brown plumage and contrasting rufous tail. **Status** Common Trinidad forest resident, usually in undergrowth. Absent from Tobago.

### Highland Hepatic Tanager *Piranga lutea faceta*                             18cm

Bulky, large-headed and bulbous-billed. Adult male has dull red head with slightly greyish ear-coverts. Underparts bright red. Upperparts duller red with brown wash to wing. Tail dark red. Upper mandible dark horn, lower paler. Legs dark grey. Female has olive-green crown. Face greenish-yellow with brighter yellow eye crescents. Underparts greenish-yellow, brighter on undertail-coverts with olive-grey flanks. Upperparts olive-green with lime fringes to secondaries, grey tertials and tail. Juvenile female like adult; juvenile male brighter, more yellow below, gradually acquires red feathers. **Voice** A loud *chup chup*, and sweeter *heep*. **SS** From Summer Tanager by more robust structure and two-toned bill. Vocalisation very different. **Status** Very uncommon resident in forested Northern Range of Trinidad, above 600m. Absent from Tobago.

### Summer Tanager *Piranga rubra rubra*                                        18cm

Adult male bright rose-red. Flight feathers and tail dusky broadly fringed red. Bill yellowish to pale horn, legs grey. Female has olive-yellow head, bright yellow underparts including undertail and olive-yellow upperparts. Wings and tail olive-grey with yellow underwing linings. Immature female recalls adult; male quickly acquires some red feathers on upperparts. **Voice** Very distinctive, rather liquid, descending *pik tchik tchik tchik*. **SS** From Highland Hepatic in all plumages by unicoloured bill. Best separated by voice. Male from Scarlet Tanager by wing and tail colour; female by occasional orange random feathering, yellow underwing linings and undertail. **Status** Rare visitor to forests on both islands, late Oct–early May.

### Scarlet Tanager *Piranga olivacea*                                          17cm

Adult breeding male has bright scarlet-red body; wings and tail black. Bill and legs pale grey. Non-breeder has bright green head, yellow-green central breast and belly, with olive rear flanks. Upperparts bright green, wings and tail blackish. Bill horn, legs grey. Female has yellowish-green head, yellow underparts, grey undertail, bright green upperparts and dusky wings and tail with pale edgings and tips. Immature recalls adult female; male has darker wings. **Voice** Usually silent. **SS** Adult breeding male unmistakable. Other plumages closely resemble Highland Hepatic and Summer Tanagers. Adult non-breeding male has blacker wings; female and immature greener upperparts, grey undertail and white underwing linings. **Status** Rare passage migrant to T&T. Northbound migration almost exclusively Apr. The only return migrants were on Tobago, early Nov 2004.

Crowned Ant Tanager

♂

♀

Summer Tanager

♂

♂
immature

♂
immature

♀

juvenile

Highland Hepatic Tanager

♂

♂
immature

♂
immature

♀
immature

Scarlet Tanager

♂
breeding

♂
moulting

♂
winter

♂
immature

♀

# NEW WORLD WARBLERS – PARULIDAE

Total of 118 species, many strongly migratory, of which 26 have been recorded on T&T; on Trinidad, three are breeding residents. During the boreal winter, another three are widespread on both islands with a fourth occurring in much smaller numbers. The rest are rare. Most have contrasting plumages associated with age, sex and season.

## Golden-winged Warbler *Vermivora chrysoptera*                                 12.5cm

Adult male has golden-yellow forecrown; hindcrown mouse-grey. Black facial wedge is narrow on lores but broadens on ear-coverts, with thin white supercilium and broad white moustachial stripe. Chin and throat black, rest of underparts grey-white. Undertail has large white oval-shaped spots. Upperparts mouse-grey, outer tail feathers white. Wing-coverts bright golden-yellow; flight feathers dark grey, fringed pale. Bill and legs black. Female has duller olive-yellow forehead and subdued face and throat markings. Immature recalls respective adult with lime-green tertial fringes. **Voice** Various *chip* or *tsip* notes. **SS** None. **Status** Very rare: three documented records for Trinidad, Dec–Jan, and an extremely late (Jun) Tobago record.

## Chestnut-sided Warbler *Dendroica pensylvanica*                                 11.5cm

Frequently droops wings and cocks tail. Adult breeding male has yellow-green crown. Face grey-white, thick black eyeline and moustachial. Broad chestnut line from moustachial along flanks, reaching thighs. Rest of underparts grey-white. Undertail has large white oval-shaped spots. Mantle and rump dirty lime-green streaked black. Tail black, fringed green. Bill and legs black. Breeding female duller. Immature and non-breeding adult have yellow-green crown, grey face with white orbital and grey-white underparts. Upperparts lime-green, wings blackish with two yellow-white wing-bars, flight feathers fringed pale. Tail black with white spots on outermost feathers. **Voice** A sharp *chip*. **SS** None. **Status** Very rare visitor to Trinidad with recent records only in Northern Range or NE forest, Dec–Apr. No records on Tobago.

## Northern Parula *Parula americana*                                 10cm

Adult male has blue-grey head with black lores and obviously broken white orbital. Throat and upper breast bright yellow with two breast-bands; upper thin and blackish; lower more diffuse and chestnut. Rest of underparts grey-white. Nape and neck-sides blue-grey. Foremantle deep lime-green, lower mantle, rump and tail blue-grey. Wing-coverts blue-grey with two white wing-bars. Flight feathers darker grey. Upper mandible black, lower yellow. Legs black with yellow feet. Female duller, lacks black lores and breast-bands. Immature recalls dull adult female but has green flight-feather fringes. **Voice** A high-pitched, *swee, chit*. **SS** From Tropical Parula by broken white orbital, black on face restricted to lores, two breast-bands, grey-white lower breast and belly, and broader wing-bars. **Status** Very rare migrant: just five documented records, four on Tobago; Nov–Feb.

## Tropical Parula *Parula pitiayumi elegans*                                 10cm

Small, plump and short-tailed. Adult male has dark blue-grey head with black lores and ear-coverts. Throat and upper breast orange, becoming bright yellow on lower breast and belly. Undertail-coverts white. Mantle greenish, rump, wings and tail deep blue with two white wing-bars and white outer tail spots. Bill two-toned, dark and pale horn. Legs orange-yellow. Female has blue-grey face without black mask and yellow not orange breast. Immature has grey-white belly and undertail-coverts, and obscure wing-bars. **Voice** Song a sweet, high-pitched trill, rising and falling in pitch; call note *chip*. **SS** See Northern Parula. **Status** Common Trinidad forest resident; abundant at Chacachacare. No acceptable records from Tobago for at least 100 years.

## Blackburnian Warbler *Dendroica fusca*                                 13cm

Adult breeding male has black crown and orange-yellow forecrown. Face bright orange with long thick black eyeline adjoining triangular lower ear-coverts patch. Throat fiery-orange. Breast yellow, becoming grey-white on belly and undertail-coverts, with broad black flanks streaking. Undertail has large white oval-shaped spots. Mantle and rump black with white tramlines. Wings black with largely white greater coverts. Tail black with extensive white on outer feathers. Bill and legs black. Non-breeding adult and immature male duller. Breeding female even duller, with grey replacing black, yellow replacing orange and two white wing-bars replacing white wing-patch. Immature female extremely dull with slight flanks streaking and two white wing-bars. **Voice** Call, *chip*. **SS** None. **Status** Rare migrant to Trinidad; just seven documented records, Jan–Apr, all in Northern Range forests. No records on Tobago.

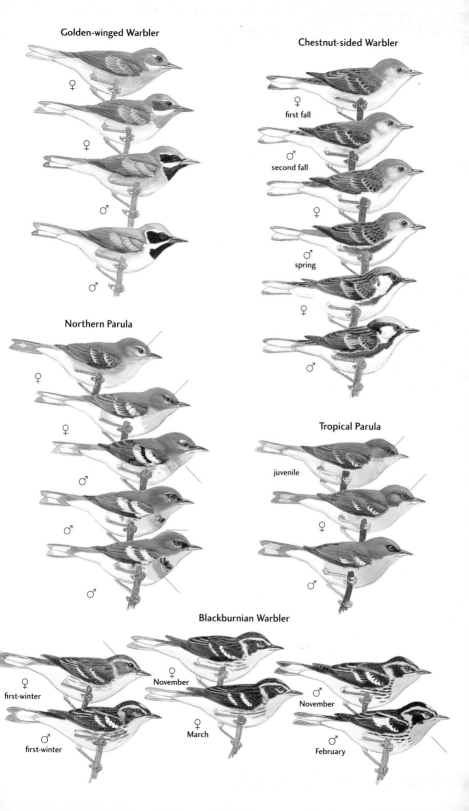

# Golden-winged Warbler

♀

♀

♂

♂

# Chestnut-sided Warbler

♀
first fall

♂
second fall

♀
spring

♂
spring

♀

♂

# Northern Parula

♀

♀

♂

♂

♂

# Tropical Parula

juvenile

♀

♂

# Blackburnian Warbler

♀
first-winter

♂
first-winter

♀
November

♀
March

♂
November

♂
February

### Yellow Warbler *Dendroica petechia aestiva*                                    12.5 cm

Some 20 or more distinct plumages of c.10 races are hypothetical for T&T, with three of the *aestiva* group most likely. Race *aestiva* is yellowish-green above; black wings and tail with broad yellowish fringes; from crown and ear-coverts to undertail-coverts intense yellow, with pale tawny streaks on breast and sides to belly. Female lacks streaks and has green to forehead. Young paler throughout, with whitish tones to face. Race *morcomi* also likely, separated by green of crown reaching to top of forehead. Males of *petechia* group all have rufous caps. **Voice** A repetitive sweet, loud *tchup* or *tchip*. **SS** Adult male unmistakable. Juvenile Bicoloured Conebill is browner above, with white undertail-coverts and long, sharp bill. **Status** Common widespread visitor to both islands, early Sep–early Apr; especially common in mangrove.

### Magnolia Warbler *Dendroica magnolia*                                    13 cm

Adult breeding male has grey crown. Face black with long white supercilium and white orbital ring. Underparts mainly bright yellow with black gorget of streaks on upper breast and heavy black flanks streaks. Undertail-coverts and central undertail-band white. Mantle black, rump yellow. Wings black with two white wing-bars and greater coverts fringes. Bill and legs black. Breeding female has grey face with faint streaking below, olive mantle streaked black and fainter white wing-bars. Non-breeding adult and immature recall dull female. **Voice** A typical *chip*. **SS** None. **Status** Accidental in T&T: just two records, Dec 1966 and Apr 1967.

### Bay-breasted Warbler *Dendroica castanea*                                    12.5 cm

In winter, essentially pale olive-green above with two white wing-bars, two white spots on undertail, underparts buffy with some chestnut on flanks. Spring female has more extensive wash to throat, whilst male develops black mask, with chestnut crown and variable amount of chestnut on throat, breast and flanks. **Voice** Usually silent. **SS** Breeding plumage unmistakable. Immature and non-breeder recall similar-age Blackpoll Warbler; best separated by (usually) unstreaked breast, olive-buff undertail-coverts, brighter upperparts, broader wing-bars and black legs. **Status** Genuinely rare, with eight documented reports from both islands, Nov–Apr.

### Blackpoll Warbler *Dendroica striata*                                    14 cm

In winter, essentially greyish-olive above with two narrow white wing-bars and white spots on undertail; pale yellowish below with very little streaking on breast and belly, but heavy streaking on breast-sides and flanks. **Voice** Occasionally a soft *chip*. **SS** See Bay-breasted Warbler. **Status** Scarce annual migrant to both islands, Oct–Nov, very occasionally to early Apr; almost all are in immature/non-breeding plumage. Found in woodland and second growth.

## Yellow Warbler

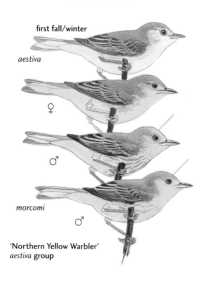

first fall/winter

*aestiva*

♀

♂

*morcomi*

♂

'Northern Yellow Warbler'
*aestiva* group

## Magnolia Warbler

♀
first fall

♂
second fall

♀
spring

♂
spring

♀

♂

## Bay-breasted Warbler

♀
first fall

♂
first fall

♀
first spring

♀          ♂
adult      first
spring     spring

♂
adult
spring

## Blackpoll Warbler

♀
first fall

♂
first fall

♀
first spring

♂
fall

♀
spring

♂
spring

## Cape May Warbler *Dendroica tigrina* 13cm

Adult male has distinctive head pattern, with black crown and eye-stripe, chestnut ear-coverts, yellow supercilium and neck collar. Yellow rump and white wing-patch; yellow below with heavy black streaking, becoming white on belly to undertail-coverts. Female similar but duller, being greyish above and paler below, with greenish rump. Younger birds lack yellow collar, but all have greenish rump and white wing-bar. **Voice** High-pitched *tsip*. **SS** Yellow rump, neck collar and lack of white tail patches prevent confusion. **Status** Very rare visitor to T&T, late Dec–early Apr. Just two records in last ten years.

## Black-throated Green Warbler *Dendroica virens* 11.5cm

Adult breeding male has olive-green crown. Face yellow with diffuse olive eyeline and dusky-olive ear-coverts. Throat and upper breast black, lower breast and belly white with bold black flanks streaking. Undertail has large white oval spots. Upperparts olive-green. Wings black with two white wing-bars. Tail blackish, outermost feathers largely white. Bill and legs black. Breeding female duller with pale throat and subdued black underparts markings. Immature and non-breeder recall breeding female. **Voice** A sharp *tsip*. **SS** None. **Status** Very rare visitor to Trinidad, with two documented records of males, Dec 1968 and Apr 1969, and two further multi-observer but as yet undocumented recent sightings.

## Prairie Warbler *Dendroica discolor paludicola* 12.5cm

Frequently flicks wings and wags tail. Adult bright olive-green above with chestnut streaks on back, blackish wings with bright yellow tips forming two wing-bars, and bright greenish-yellow fringes. Face pattern distinctive, comprising broad eyebrow, crescent below eyes and dark crescent on neck-sides. Male is more richly coloured and heavily streaked on flanks. Juvenile duller, and greyish-green above. Almost entire undertail is white. **Voice** Call a typical *tchip*. **SS** Facial pattern and flank streaking makes confusion unlikely. **Status** Accidental to Trinidad: immature Mar 1978. Prefers dry scrub and second growth.

## Yellow-rumped Warbler *Dendroica coronata coronata* 14cm

Brown to greyish-brown above in all plumages, except breeding male, and all except juvenile have two white or buffy wing-bars, and all possess yellow rump. Underparts dingy white, streaked brown on breast and flanks. All but juvenile have broad line of yellow on body-sides. Breeding male is blue above with black centres to feathers, a bold yellow coronal streak, white eyebrow and crescent below eye, bold black face-sides and streaks on breast and belly-sides. Juvenile almost uniform above. **Voice** A sharp *chek*. **SS** Yellow on breast-sides and rump distinctive in all plumages. **Status** Accidental on Tobago, with just two documented records, both in early months of year.

## Black-throated Blue Warbler *Dendroica caerulescens cairnsi* 11.5cm

Adult male blue above, black wings with very broad blue fringes and white patch on primaries; black face and throat, with long broad line reaching flanks, rest of underparts white; undertail dark with bold white disc on each outer feather. Female brownish-olive above, with clean white eyebrow and small crescent below eye, small white wing spot; buffy below. Juvenile duller – both sexes have very weak pale area on undertail. *D. c. caerulescens* (hypothetical) is brighter above, with clear blue crown and upperparts, and smaller white patch on wing. **Voice** Soft dry *tic*. **SS** Adult male unmistakable. Combination of white eye-stripe and primary patch separates female from other warblers. **Status** Accidental Trinidad: adult males, Northern Range, Mar 1966 and 1992.

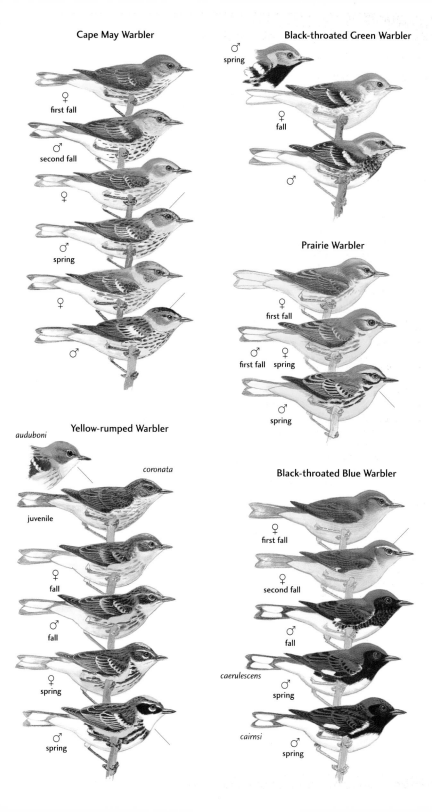

## Cape May Warbler

♀
first fall

♂
second fall

♀

♂
spring

♀

♂

## Black-throated Green Warbler

♂
spring

fall

♂

## Prairie Warbler

♀
first fall

♂          ♀
first fall   spring

♂
spring

## Yellow-rumped Warbler

*auduboni*

*coronata*

juvenile

♀
fall

♂
fall

♀
spring

♂
spring

## Black-throated Blue Warbler

♀
first fall

♀
second fall

♂
fall

*caerulescens*
♂
spring

*cairnsi*
♂
spring

## Cerulean Warbler *Dendroica cerulea* 12cm

Adult breeding male has bright blue-grey crown with duller face and black eyeline. Throat and neck-sides white, bordered by thin dusky upper-breast band. Rest of underparts white with diffuse black flanks streaking. Undertail has large white oval spots. Upperparts blue-grey with indistinct black streaks. Wings black with two white wing-bars and blue fringes to flight feathers. Tail dark blue with white blocks in outer feathers, visible when fanned. Bill and legs black. Non-breeder has indistinct white supercilium and incomplete breast-band. Breeding female has dull blue-green head with yellow-white supercilium and black eyeline. Underparts grey-white with yellow tone to upper breast and faint flanks streaking. Upperparts greener than male. Juvenile and non-breeding female have greener mantle and brighter yellow underparts. **Voice** A typical *chip*. **SS** Adult unmistakable. Immature from other warblers by bright bluish plumage and bold wing-bars. **Status** Accidental on Trinidad, one in Northern Range, Nov 2000.

## Golden-crowned Warbler *Basileuterus culicivorus olivascens* 12.5cm

Chunky with sturdy legs. Adult has grey head with thin, often obscured, dull orange central crown-stripe, broad black coronal bands, black lores and long creamy white supercilium. Breast and belly sulphur-yellow, becoming lemon on undertail-coverts. Upperparts olive-grey-brown. Bill dark horn, legs straw-yellow. Juvenile duller, lacks head stripes and greyish-olive above with two indistinct buff wing-bars; olive breast and dull yellow belly and undertail-coverts. **Voice** Often flicks tail and droops wings when uttering sharp *stik* alarm-call. Song a 6–7 note fluty, descending, *whee whee whee*. **SS** None. **Status** Common and widespread Trinidad resident, favouring forest, second growth and cultivated estates. Absent on Tobago.

## Prothonotary Warbler *Protonotaria citrea* 14cm

Bulky and large-headed. Adult male has golden-yellow head, brightest on forecrown, with black eyes. Throat and upper breast golden-yellow, paler on lower breast and belly. Undertail-coverts white. Undertail has large white oval spots. Mantle olive-green, wings, rump and tail blue-grey with white blocks in outer tail. Bill dark above, pale below; legs grey. Female lacks richest golden tones to head and upper breast. Juvenile duller. **Voice** Loud and strident, *tsip*. **SS** None. **Status** Uncommon visitor to both islands, late Oct–late Mar. Usually found close to water, preferring mangroves and their edges.

## Black-and-white Warbler *Mniotilta varia* 13cm

Nuthatch-like feeding behaviour. Adult breeding male has black head with white central crown-stripe, broad supercilium flaring over eye, moustachial stripe and thin orbital. Throat black. Rest of underparts white with black breast and flanks streaking, and spotted undertail-coverts. Undertail has large white oval spots. Upperparts striped black and white. Tail black. Wings black with two white wing-bars, white tertial tips and pale flight-feather fringes. Bill and legs grey. Non-breeding male duller with mottled throat. Female has grey-white ear-coverts, lores and throat. Underparts white with faint flanks streaking. Juvenile recalls adult female; male more boldly marked, female slightly duller. **Voice** Normally silent. **SS** None. **Status** Rare winter visitor to forest on both islands, mostly late Oct–mid-Apr.

## American Redstart *Setophaga ruticilla* 11.5cm

Frequently droops wings while spreading tail. Adult male has black head, throat and upper breast. Belly and undertail-coverts grey-white with bright orange pectoral patch. Undertail duller orange with broad black terminal bar. Upperparts black with broad, bright orange wing-bar and orange outer tail feathers with black terminal bar. Bill and legs black. Female has grey head with thin white orbital and faint white loral stripe. Underparts white with bright yellow pectoral patch. Upperparts dark olive with yellow wing-bar. Tail dusky with yellow bases to outer feathers. Immature male recalls adult female with orange pectoral patch. **Voice** High-pitched *chip*, less strident than Yellow Warbler. **SS** None. **Status** Common and widespread winter visitor to T&T, Sep– early Apr.

## Golden-crowned Warbler

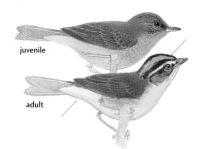

juvenile

adult

## Cerulean Warbler

first fall

♀
spring

♂
fall

♂
spring

## Prothonotary Warbler

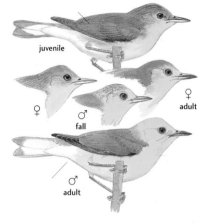

juvenile

♀

♂
fall

♀
adult

♂
adult

## Black-and-white Warbler

♀
first fall

♀
spring adult

♂
first fall

♂
adult fall

♂
spring adult

## American Redstart

juvenile

♀

♂
immature

♂

♂

## Northern Waterthrush *Seiurus noveboracensis*                    12.5cm

Horizontal gait and perpetually bobbing tail. Adult has dark chocolate-brown head with long creamy supercilium narrowing behind eye, and greyer cheeks. Underparts yellow-white densely streaked dark brown, streaks blurred on rear flanks. Upperparts dark chocolate-brown. Bill thin and dark horn. Legs brownish-pink. Juvenile recalls adult, with pale tertial fringes. **Voice** Distinct metallic *tink*. **SS** Louisiana Waterthrush *S. motacilla* (unrecorded in T&T) has supercilium flaring broadly on neck-sides, unmarked throat and whiter underparts with buff flanks. **Status** Common winter visitor in T&T, mostly late Sep– early May. Favours forest and mangrove edge, usually feeding near ground, often on quiet roadsides.

## Tennessee Warbler *Vermivora peregrina*                    12cm

Adult breeding male has pale grey head with black eyeline and white supercilium. Olive-green above, off-white below. Female uniform olive-green above, lacking any grey, and creamy-white below. Juvenile has two yellowish wing-bars, and is more yellowish below. **Voice** Thin high-pitched *ptzee*. **SS** Adult unmistakable; immature from Yellow Warbler by white undertail-coverts. **Status** Very rare migrant. Several undocumented but probably correct reports in Northern Range, Trinidad, in winter.

## Ovenbird *Seiurus aurocapilla*                    12.5cm

Olive head and upperparts with prominent white orbital, broad, dull orange coronal stripe bordered with black. White below, with black arrowhead streaks on mesial, breast and flanks. **Voice** Hard resonant *chip*. **SS** None. **Status** Accidental in T&T: just three documented records Nov–Mar and none since 1971. Walks with continual bobbing action.

## Common Yellowthroat *Geothlypis trichas trichas*                    12.5cm

Olive-green above, yellow below, buffy on sides and flanks, with whitish on belly and undertail-coverts. Adult male has broad black mask from forehead, tapering on neck-sides, distinctively edged pale grey above. Female lacks mask, has diffuse buffy eyebrow and is browner olive above. Juvenile like female but young male gradually develops mask, with pale fringe acquired last. **Voice** A fine *stit* or *chip*. **SS** Adult male from Masked Yellowthroat by olive crown, narrow white border to mask and pale buff belly and flanks. Female by browner crown, duller upperparts and contrasting underparts. **Status** Accidental Trinidad: three records, most recent Mar 1965.

## Masked Yellowthroat *Geothlypis aequinoctialis*                    12.5cm

Male olive-green above, bluish-grey on forecrown, with black mask from forehead ending square-cut on ear-coverts; entirely rich yellow below, washed green on sides and flanks. Female browner olive above, lacks mask, and is paler on face and eyebrow; washed olive on sides and flanks. Juvenile slightly paler above, pale buffy on face, sides and flanks, washed-out creamy below. **Voice** Song a fluty descending *weechew weechew*, 8–15 notes; call a disyllabic *whee hu*. **SS** See Common Yellowthroat. **Status** Locally uncommon resident of freshwater swamps and lowland pastures on Trinidad. Absent from Tobago.

## Hooded Warbler *Wilsonia citrina*                    13cm

Terrestrial; often flicks tail. Yellowish-olive above, with large bright yellow mask, boldly fringed black, becoming full bib on throat, and male has black extending to nape. Bright yellow below. Juvenile like female but lacks black 'frame' and has yellow face washed greenish. **Voice** Typical *tchip*. **SS** Adult male unmistakable. Female separated from any other warbler by extensive white in outer tail. **Status** Accidental Trinidad: adult male Dec 1978; immature female Jan 2006.

## Canada Warbler *Wilsonia canadensis*                    13cm

Uniform grey above, with yellow supraloral and orbital ring, and black lores (all plumages); underparts yellow, with white undertail-coverts. Young may show emerging blackish streaks on breast, female usually has well-streaked breast, male has broad streaks on breast, also on lower cheeks and forecrown. **Voice** Quiet *chut*. **SS** None. **Status** Accidental Trinidad: male, Northern Range, Dec 1993.

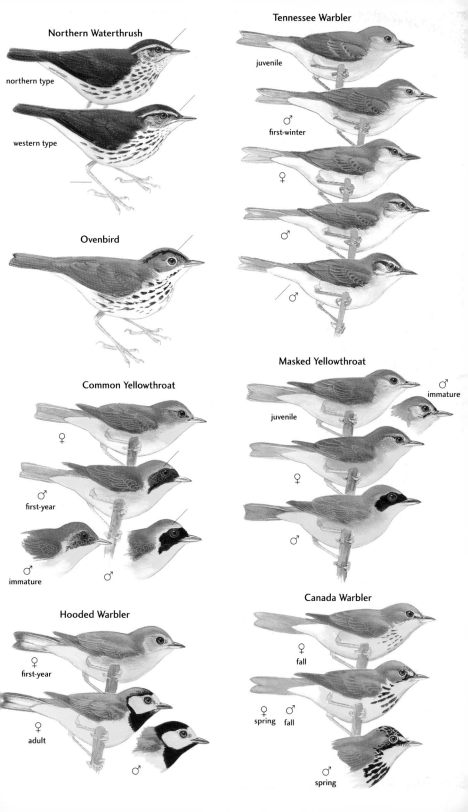

Northern Waterthrush

northern type

western type

Tennessee Warbler

juvenile

♂
first-winter

♀

♂

♂

Ovenbird

Common Yellowthroat

♀

♂
first-year

♂
immature

♂

Masked Yellowthroat

♂
immature

juvenile

♀

♂

Hooded Warbler

♀
first-year

♀
adult

♂

Canada Warbler

♀
fall

♀      ♂
spring  fall

♂
spring

# NEW WORLD BLACKBIRDS – ICTERIDAE

Some 101 species occur throughout the New World. Just three species of cowbirds and grackles occur in T&T, variable in size, gregarious and noisy in character. Six orioles and oropendolas have been recorded in T&T; one is resident on both islands, three on Trinidad alone, and two are rare visitors. All share yellow or orange and black plumage. Three species of New World blackbird have occurred on Trinidad. Two are resident, the third an accidental visitor. Just one occurs, rarely, on Tobago.

### Crested Oropendola *Psarocolius decumanus insularis* 33–43cm

Adult male has sooty-black head with blue eyes. Black crest rarely visible. Underparts black with chestnut-brown undertail-coverts. Nape black, grading to deep chestnut on lower rump/uppertail-coverts. Tail long, bright lemon-yellow with two central feathers chestnut-brown tipped yellow. Bill long, conical and cream. Legs black. Female smaller, shorter tailed and dusky not black. Immature recalls adult. **Voice** Loud chirpy *kiop* or *querp*. Male display-call a unique series of bubbles, gurgles and trills. **SS** From Yellow-rumped Cacique by much larger size, longer tail, all-black wings, chestnut-brown undertail-coverts and rump, and mainly yellow tail. **Status** Very common and widespread resident on both islands.

### Yellow-rumped Cacique *Cacicus cela cela* 25–29cm

Adult male has black head with short scraggy crest and blue eyes. Breast and belly black, undertail-coverts bright yellow. Nape and fore mantle black. Rest of upperparts and basal half of tail bright yellow; tail distally black. Wings black with yellow wedge on inner coverts. Bill cream, legs black. Female smaller and duller. Juvenile recalls adult but duller with brown eyes. **Voice** Wide-ranging repertoire, from haunting fluty notes to harsh chattering. Often imitates other species. **SS** From Crested Oropendola, with which it often associates, by smaller size, shorter thinner bill, shorter black tail and yellow rump. **Status** Common resident in lowland C, E and N Trinidad. Absent from Tobago.

### Moriche Oriole *Icterus chrysocephalus* 23cm

Rather long-tailed. Adult black except golden-yellow crown, nape, lesser coverts and rump. Bill quite long, slender, decurved and black; legs black with dull yellow thighs only visible at close range. Juvenile like brownish adult. **Voice** Single, rather harsh *kraak* and repetitive *heep heep*; song a series of sweet undulating musical whistles. **SS** None. **Status** Very scarce Trinidad resident, restricted to C and E savannas with Moriche palms. Absent from Tobago.

### Yellow-hooded Blackbird *Chrysomus icterocephalus icterocephalus* 19cm

Adult male has bright yellow head with black chin and lores. Throat and upper breast yellow; remaining plumage black. Bill and legs black. Female and immature much duller. Most have dark olive heads with long yellow supercilium surrounding olive-green ear-coverts. Chin, throat and upper breast variably yellow; lower breast olive-yellow, becoming grey-brown on belly and undertail-coverts. Upperparts olive-brown with black feather centres. Tail black. Wings dark grey fringed yellow, tertials black. **Voice** Long drawn-out *tzweee* and varied short stutters, wheezes and clucks. **SS** None. **Status** Very common Trinidad resident, in wet pastures, rice fields, freshwater marshes and coastal mangroves. Absent from Tobago.

**Crested Oropendola**

**Yellow-rumped Cacique**

adult

juvenile

**Moriche Oriole**

juvenile

adult

**Yellow-hooded Blackbird**

juvenile

♀

♂

♂

### Yellow Oriole *Icterus nigrogularis trinitatis* 20cm

Adult has golden-yellow head with black lores and mask. Chin, throat and neck-sides black, breast, belly and undertail-coverts bright yellow, undertail black. Nape and fore mantle golden-yellow lightly flecked green, lower mantle and rump slightly duller. Tail black. Wings black with greater coverts and tertials fringed white. Bill black, legs blue-grey. Juvenile has reduced black bib, duller yellow underparts, greener mantle and brown wings and tail with buff wing-bars. **Voice** Highly variable. A double note *stup-seeer* and various *chuip chee-u-ip* calls, a harsh *kraa* or *cark* and repetitive *wicu wicu wicu*. **SS** None. **Status** Common widespread Trinidad resident. Absent from Tobago.

### Orchard Oriole *Icterus spurius spurius* 18cm

Adult male has black head, throat and upper breast, rest of underparts deep chestnut-red. Upperparts and tail black, rump chestnut. Inner wing-coverts chestnut, greater coverts dusky fringed white. Flight feathers black narrowly fringed white, underwing-coverts chestnut. Bill blue-grey tipped black, legs darker. Adult female has olive-yellow head and nape, brightest around eye. Underparts bright greenish-yellow. Mantle grey, rump greenish-yellow, tail duller olive. Upperwing dark grey with two white wing-bars and flight-feather fringes. Immature male like adult female with black lores and throat. **Voice** Sharp *chek* or *cluck*. **SS** Adult unmistakable. Immature from Baltimore Oriole by greenish-yellow underparts and duller wing-bars. **Status** Accidental on Trinidad: male at Nariva, Dec 2006.

### Baltimore Oriole *Icterus galbula galbula* 17cm

Adult male has black head and throat, rest of underparts bright orange with black base to undertail-coverts. Nape and fore mantle black, lower mantle and rump orange. Uppertail-coverts and central tail feathers black, rest orange. Wings black with orange lesser coverts, white tips to tertials, median and greater coverts, and white fringes to flight feathers. Bill and legs blue-grey. Subadult male and adult female duller and paler, with dark feathers fringed buff and two white wing-bars. Subadult female has less black on head, yellow chin and throat, and three wing-bars. **Voice** A harsh rattle; in flight a metallic *szeeet*. **SS** None. **Status** Very rare migrant: handful of sightings in T&T, Jan–Apr. Prefers cultivated estates and forest edge.

### Shiny Cowbird *Molothrus bonariensis minimus* 19cm

Adult male flat-crowned and black with iridescent purple gloss. Bill conical and steel grey; legs black. Female has dark grey head with paler supercilium and black eyes. Underparts pale grey-brown with diffuse mottling. Upperparts dusky, wings browner with paler covert fringes. Juvenile male recalls adult female with indistinct yellowish supercilium and few shiny black feathers. **Voice** Song a pleasing variety of musical whistles and trills. Series of full-toned *cluck* and *tchip* notes. **SS** Male from White-lined Tanager and Carib Grackle by flat crown, glossy sheen, black eyes and square-ended tail. **Status** Very common widespread resident in T&T.

### Giant Cowbird *Molothrus oryzivorus oryzivorus* 28–36cm

Large and long-tailed with undulating flight. On perch looks bull-necked. Adult male black with bronze-purple sheen and ruffed appearance to neck. Eyes red. Bill and legs black. Female smaller and lacks metallic sheen, with yellower eyes. Immature recalls adult female but browner with pale bill and whitish eyes. **Voice** Loud, sharp series of whistles, squeaks and chatters. **SS** None. **Status** Locally common resident: on Trinidad prefers open savannas and pastures; on Tobago found on scrubby hillsides, especially in NE.

**Yellow Oriole**

juvenile

♀

♂

**Orchard Oriole**

♀

juvenile

♂

immature

**Shiny Cowbird**

juvenile

♀

♂

♂

**Baltimore Oriole**

juvenile

juvenile

♀

♂

♀

♂

♀

**Giant Cowbird**

♂

# PLATE 106: NEW WORLD BLACKBIRDS III

## Carib Grackle *Quiscalus lugubris lugubris*     23–27cm
Long, wedge-shaped tail. Adult male glossy blue-black with yellow-white eyes. Bill black and decurved; legs black. Female has flat tail and is dark brown above, without any gloss, paler below, especially on throat. Juvenile female similar but has brown eyes. Juvenile male also like female, but has patches of blue gloss. **Voice** Rattles, chicks and clucks interspersed with full ringing notes and shrill whistles. Frequently utters metallic *hwenk....hwenk*. **SS** From Shiny Cowbird and male White-lined Tanager by wedge-shaped tail and pale eyes. **Status** Abundant and widespread resident in lowlands of T&T.

## Red-winged Blackbird *Agelaius phoeniceus*     23cm
Adult male all black with bold red lesser coverts and (less distinct) yellow median coverts. Red appears as 'headlight' in flight. Bill and legs black. Subadult male has buff feather fringes, black innermost lesser coverts and basally black median coverts. Female has dark brown head with buff supercilium, black eyeline, buff moustachial and black malar. Throat buff, rest of underparts grey-white heavily streaked black. Upperparts dark brown fringed reddish. Flight feathers and tail black. Juvenile recalls adult female with horn-coloured bill. **Voice** Series of liquid notes followed by gurgling *kor law wee*, and buzzy descending *zeeeer*. **SS** None. **Status** Accidental on Trinidad: one near Caroni Swamp, 1980 (subspecies unknown).

## Red-breasted Blackbird *Sturnella militaris militaris*     18cm
Short-winged, short-tailed and plump-bodied. Adult male has black head and neck-sides. Throat, breast and upper belly bright red, rest of underparts black. Upperparts black with pale tips to coverts and fringes to flight feathers. Red shoulder-patch only conspicuous in flight. Bill and legs black. Female has dark brown crown with buff central stripe and broad creamy supercilium. Face buff with thin black line. Underparts buff with rose-red wash to breast and variable dark streaking on neck-sides and flanks. Mantle and rump dark brown with buff tramlines. Tail dark brown barred grey. Wings dark grey-brown with tertials fringed buff. Bill horn. Juvenile recalls adult female but paler and lacks red on underparts. **Voice** Adult male utters dry *che wit a whzeeeee* in parachute display-flight. A hard *chip*. **SS** Adult unlikely to be confused. Juvenile from non-breeding Bobolink by larger size, buff-brown plumage and lack of white orbital. **Status** Common resident, widespread in lowland grasslands, pastures and water meadows, even urban parks on Trinidad. Rare on Tobago.

## Bobolink *Dolichonyx oryzivorus*     15cm
Adult breeding male has black forecrown, face and underparts. Rear crown and nape yellow-buff. Mantle black, rump and uppertail-coverts white. Wings and tail black with large white scapular patch. In all other plumages, crown black with cream central stripe. Face yellow-buff with broad cream supercilium, thin black eyeline and white orbital ring. Underparts yellow-buff with narrow black rear-flanks streaks. Mantle and rump olive-brown streaked black, bordered by cream scapular lines. Tail grey with distinctly pointed tips. Wings black with coverts broadly fringed buff. Bill conical and dull pink; legs reddish-brown. **Voice** A single *pink*. **SS** Adult breeding male unmistakable. Other plumages from Red-breasted Blackbird by yellower plumage, white orbital and pinkish bill. **Status** Boreal passage migrant to South America, but only a rare visitor to T&T, favouring wet lowlands; mostly in Oct.

# WAXBILLS AND ALLIES – ESTRILDIDAE
Waxbills are an Old World family, of African origin. They are popular cagebirds and escapes have resulted in many species establishing feral populations around the world. Only one species is present on Trinidad.

## Common Waxbill *Estrilda astrild* (subspecies unknown)     11cm
Olive-brown above, faintly and finely barred dusky, broad red line through eyes, face-sides pale buffy, rest of underparts buffy with red flush to centre of belly, and black undertail-coverts. Bill red. Juvenile similar but has dark bill, lacks any red on belly, and has olive-brown undertail-coverts. **Voice** Loud, rather explosive, high-pitched *cheep*. **SS** None. **Status** Introduced on Trinidad in 1990, with self-sustaining population in C and W Trinidad, in waterside reeds and fringes. One recorded on Tobago, April 2007.

**Carib Grackle**

♀

♂

**Red-winged Blackbird**

juvenile

♀

immature ♂

♂
fresh winter
plumage

**Red-breasted Blackbird**

juvenile

♀

♂

♂

**Bobolink**

♀

♂

♂

**Common Waxbill**

juvenile

adult

## SISKINS – SUBFAMILY CARDUELINAE

Only two species have occurred in T&T and, of these, Red Siskin is now extirpated due to relentless trapping for the illegal cagebird trade. Lesser Goldfinch is seriously threatened.

### Red Siskin *Carduelis cucullata* 11.5cm

Adult male blood red, with black head, wings and tail; wings have bold white fringes to tertials, red tips to coverts and a large reddish-pink patch on flight feathers. Female soft warm grey, darker above, with reddish flush to rump and breast, and black wings and tail (pattern largely as male). Young similar to female above, but much duller, and buffy below. **Voice** Contact-call a rough *djut djut*, song a clear ringing high-pitched *ka-hee*, with various twitters and trills. **SS** None. **Status** Locally extirpated on Trinidad and globally threatened. Last documented sighting in Arima Valley, 1960.

### Lesser Goldfinch *Carduelis psaltria columbiana* 10cm

Adult male black above, with large white wing-panel and underside to outer rectrices. In fresh plumage, tertials have large white corners (lost with wear); underparts pure lemon-yellow. Female soft olive-green above, with black wings and white fringes to tertials (lost with wear), and yellow underparts lightly washed olive-green. Juvenile pale brown above, soft yellowish below. **Voice** Plaintive *pee-eee*. **SS** Adult male unmistakable. Female/immature from Grassland Yellow Finch by plainer face, unstreaked upperparts and white wing-bars. **Status** Nine at Mt St Benedict, Trinidad, Feb 2005.

## EUPHONIAS – SUBFAMILY EUPHONIINAE

Small, generally dumpy, short-tailed and stubby-billed. Adult males dark blue and yellow; species identification relies on throat, nape and rump colour. Female generally olive-green. Three species theoretically occur on Trinidad, but one is probably locally extirpated. Just one occurs on Tobago.

### Trinidad Euphonia *Euphonia trinitatis* 11.5cm

Adult male has crown golden-yellow, rest of head to upper breast and upperparts rich deep blue. Underparts golden-yellow, base of undertail white. Female/juvenile olive-green above, with yellowish chin; cheeks, throat, breast-sides and flanks to undertail-coverts pure yellow; breast and upper belly greyish-white. **Voice** Far-carrying, doubled *pee pee*, often repeated 12–15 times. Single, rather mournful *siu*. **SS** Male from Violaceous Euphonia by blue throat and lack of white in flight feathers. Female by grey band on lower breast. **Status** Uncommon yet widespread resident on Trinidad, favouring open scrubby hillsides and hill forest, especially with mistletoe, and mangrove on W coast. One old record on Tobago of suspect origin.

### Violaceous Euphonia *Euphonia violacea rodwayi* 11.5cm

Adult male deep violaceous blue from forecrown to tail; forehead, lores and chin to breast and sides rich golden-orange, gradually becoming more yellow, with yellow undertail-coverts; most of undertail white. Female olive-green above, pale olive-yellow below, with yellowish undertail-coverts. **Voice** Mimics many species. Song very musical; calls harsh and strident, frequently a sharp *chit a chit chit*. **SS** Juvenile Golden-rumped larger and uniformly darker below. **Status** Common and widespread Trinidad resident, in estates, second growth and forest. Scarce on Tobago, where restricted to hill forest.

### Golden-rumped Euphonia *Euphonia cyanocephala cyanocephala* 12.5cm

Adult male has forehead and face to throat black, forecrown to nape bright sky blue, back, wings and tail black with purple gloss; rump and short uppertail-coverts and breast to undertail-coverts bright golden-yellow. Female has yellow to orange forehead, blue forecrown to nape, and rest of plumage olive-green, yellowish below. Juvenile virtually uniform olive-green. **Voice** A musical chirp. **SS** Adult male unmistakable. Female from other euphonias by pale blue nape and yellow forehead. **Status** Very rare, probably no longer resident on Trinidad (no recent records). Most former sightings from hill forest. Absent from Tobago.

## Red Siskin

juvenile

immature

♂

♀

♂

## Lesser Goldfinch

juvenile

juvenile
first moult

fresh
♀

worn
♂

fresh
♂

## Trinidad Euphonia

## Violaceous Euphonia

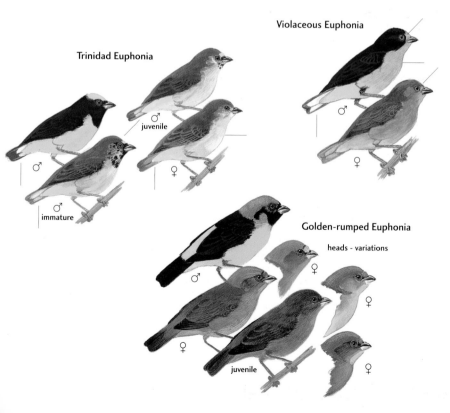

juvenile
♂

♂

♀

♂

immature

♂

♀

## Golden-rumped Euphonia

heads - variations

♂

♀

♀

♀

♀

juvenile

♀

# OFFICIAL CHECKLIST OF THE BIRDS OF
# TRINIDAD & TOBAGO

The official list of species of Trinidad & Tobago stands at 467 as at May 2007. This list is updated periodically to reflect additions brought about by the significant increase in coverage by both local and visiting birdwatchers.

This list has two components: (a) species adjudged as correctly identified by the Trinidad & Tobago Rare Birds Committee, since September 1995, and that occur naturally or have a sustainable breeding population; and (b) species adjudged acceptable by Richard ffrench prior to the TTRBC's inauguration.

Nomenclature and sequence mainly follow those of the American Ornithologists Union (AOU) and their South American Checklist Committee (SACC).

## Tinamous – Tinamidae
❏ Little Tinamou *Crypturellus soui*

## Screamers – Anhimidae
❏ Horned Screamer *Anhima cornuta*

## Ducks and Geese – Anatidae
❏ White-faced Whistling Duck *Dendrocygna viduata*
❏ Black-bellied Whistling Duck *Dendrocygna autumnalis*
❏ Fulvous Whistling Duck *Dendrocygna bicolor*
❏ Snow Goose *Chen caerulescens*
❏ Muscovy Duck *Cairina moschata*
❏ Comb Duck *Sarkidiornis melanotos*
❏ American Wigeon *Anas americana*
❏ Green-winged Teal *Anas carolinensis*
❏ Northern Pintail *Anas acuta*
❏ White-cheeked Pintail *Anas bahamensis*
❏ Blue-winged Teal *Anas discors*
❏ Northern Shoveler *Anas clypeata*
❏ Southern Pochard *Netta erythrophthalma*
❏ Ring-necked Duck *Aythya collaris*
❏ Lesser Scaup *Aythya affinis*
❏ Masked Duck *Nomonyx dominicus*

## Chachalacas and Guans – Cracidae
❏ Rufous-vented Chachalaca *Ortalis ruficauda*
❏ Trinidad Piping Guan *Pipile pipile*

## Grebes – Podicipedidae
❏ Least Grebe *Tachybaptus dominicus*
❏ Pied-billed Grebe *Podilymbus podiceps*

## Petrels and Shearwaters – Procellariidae
❏ Bulwer's Petrel *Bulweria bulwerii*
❏ Cory's Shearwater *Calonectris diomedea*
❏ Sooty Shearwater *Puffinus griseus*
❏ Great Shearwater *Puffinus gravis*
❏ Manx Shearwater *Puffinus puffinus*
❏ Audubon's Shearwater *Puffinus lherminieri*

## Storm-Petrels – Hydrobatidae
❏ Wilson's Storm-Petrel *Oceanites oceanicus*
❏ Leach's Storm-Petrel *Oceanodroma leucorhoa*

## Tropicbirds – Phaethontidae
❏ Red-billed Tropicbird *Phaethon aethereus*
❏ White-tailed Tropicbird *Phaethon lepturus*

## Pelicans – Pelicanidae
❏ Brown Pelican *Pelicanus occidentalis*

## Boobies – Sulidae
❏ Masked Booby *Sula dactylatra*
❏ Red-footed Booby *Sula sula*
❏ Brown Booby *Sula leucogaster*
❏ Northern Gannet *Morus bassanus*

## Cormorants – Phalacrocoracidae
❏ Neotropic Cormorant *Phalacrocorax brasilianus*

## Anhinga – Anhingidae
❏ Anhinga *Anhinga anhinga*

## Frigatebirds – Fregatidae
❏ Magnificent Frigatebird *Fregata magnificens*

## Herons – Ardeidae
❏ Rufescent Tiger Heron *Tigrisoma lineatum*
❏ Agami Heron *Agamia agami*
❏ Boat-billed Heron *Cochlearius cochlearius*
❏ Pinnated Bittern *Botaurus pinnatus*
❏ Least Bittern *Ixobrychus exilis*
❏ Stripe-backed Bittern *Ixobrychus involucris*
❏ Black-crowned Night Heron *Nycticorax nycticorax*
❏ Yellow-crowned Night Heron *Nyctanassa violacea*
❏ Green Heron *Butorides virescens*
❏ Striated Heron *Butorides striata*
❏ Cattle Egret *Bubulcus ibis*
❏ Grey Heron *Ardea cinerea*
❏ Purple Heron *Ardea purpurea*

- ❏ Great Blue Heron *Ardea herodias*
- ❏ Cocoi Heron *Ardea cocoi*
- ❏ Great Egret *Ardea alba*
- ❏ Tricoloured Heron *Egretta tricolor*
- ❏ Reddish Egret *Egretta rufescens*
- ❏ Western Reef Heron *Egretta gularis*
- ❏ Little Egret *Egretta garzetta*
- ❏ Snowy Egret *Egretta thula*
- ❏ Little Blue Heron *Egretta caerulea*

## Ibises and Spoonbills – Threskiornithidae
- ❏ White Ibis *Eudocimus albus*
- ❏ Scarlet Ibis *Eudocimus ruber*
- ❏ Glossy Ibis *Plegadis falcinellus*
- ❏ Eurasian Spoonbill *Platalea leucorodia*
- ❏ Roseate Spoonbill *Platalea ajaja*

## Storks – Ciconiidae
- ❏ Maguari Stork *Ciconia maguari*
- ❏ Jabiru *Jabiru mycteria*
- ❏ Wood Stork *Mycteria americana*

## New World Vultures – Cathartidae
- ❏ Turkey Vulture *Cathartes aura*
- ❏ Black Vulture *Coragyps atratus*
- ❏ King Vulture *Sarcoramphus papa*

## Flamingos – Phoenicopteridae
- ❏ Greater Flamingo *Phoenicopterus ruber*

## Osprey – Pandionidae
- ❏ Osprey *Pandion haliaetus*

## Hawks, Kites and Eagles – Accipitridae
- ❏ Grey-headed Kite *Leptodon cayanensis*
- ❏ Hook-billed Kite *Chondrohierax uncinatus*
- ❏ Swallow-tailed Kite *Elanoides forficatus*
- ❏ Pearl Kite *Gampsonyx swainsonii*
- ❏ White-tailed Kite *Elanus leucurus*
- ❏ Snail Kite *Rostrhamus sociabilis*
- ❏ Double-toothed Kite *Harpagus bidentatus*
- ❏ Plumbeous Kite *Ictinia plumbea*
- ❏ Long-winged Harrier *Circus buffoni*
- ❏ Crane Hawk *Geranospiza caerulescens*
- ❏ White Hawk *Leucopternis albicollis*
- ❏ Grey-lined Hawk *Asturina nitida*
- ❏ Common Black Hawk *Buteogallus anthracinus*
- ❏ Rufous Crab Hawk *Buteogallus aequinoctialis*
- ❏ Great Black Hawk *Buteogallus urubitinga*
- ❏ Savanna Hawk *Buteogallus meridionalis*
- ❏ Black-collared Hawk *Busarellus nigricollis*
- ❏ Broad-winged Hawk *Buteo platypterus*
- ❏ Short-tailed Hawk *Buteo brachyurus*
- ❏ Swainson's Hawk *Buteo swainsoni*
- ❏ White-tailed Hawk *Buteo albicaudatus*

- ❏ Zone-tailed Hawk *Buteo albonotatus*
- ❏ Black Hawk-Eagle *Spizaetus tyrannus*
- ❏ Ornate Hawk-Eagle *Spizaetus ornatus*

## Caracaras and Falcons – Falconidae
- ❏ Northern Crested Caracara *Caracara cheriway*
- ❏ Yellow-headed Caracara *Milvago chimachima*
- ❏ Common Kestrel *Falco tinnunculus*
- ❏ American Kestrel *Falco sparverius*
- ❏ Merlin *Falco columbarius*
- ❏ Bat Falcon *Falco rufigularis*
- ❏ Orange-breasted Falcon *Falco deiroleucus*
- ❏ Aplomado Falcon *Falco femoralis*
- ❏ Peregrine *Falco peregrinus*

## Limpkin – Aramidae
- ❏ Limpkin *Aramus guarauna*

## Rails, Crakes and Gallinules – Rallidae
- ❏ Clapper Rail *Rallus longirostris*
- ❏ Grey-necked Wood Rail *Aramides cajanea*
- ❏ Rufous-necked Wood Rail *Aramides axillaris*
- ❏ Grey-breasted Crake *Laterallus exilis*
- ❏ Yellow-breasted Crake *Porzana flaviventer*
- ❏ Ash-throated Crake *Porzana albicollis*
- ❏ Sora *Porzana carolina*
- ❏ Paint-billed Crake *Neocrex erythrops*
- ❏ Spotted Rail *Pardirallus maculatus*
- ❏ Common Moorhen *Gallinula chloropus*
- ❏ Purple Gallinule *Porphyrio martinica*
- ❏ Azure Gallinule *Porphyrio flavirostris*
- ❏ Caribbean Coot *Fulica caribaea*
- ❏ American Coot *Fulica americana*

## Sungrebe – Heliornithidae
- ❏ Sungrebe *Heliornis fulica*

## Lapwings and Plovers – Charadriidae
- ❏ Pied Lapwing *Vanellus cayanus*
- ❏ Southern Lapwing *Vanellus chilensis*
- ❏ American Golden Plover *Pluvialis dominica*
- ❏ Black-bellied Plover *Pluvialis squatarola*
- ❏ Semipalmated Plover *Charadrius semipalmatus*
- ❏ Wilson's Plover *Charadrius wilsonia*
- ❏ Killdeer *Charadrius vociferus*
- ❏ Snowy Plover *Charadrius alexandrinus*
- ❏ Collared Plover *Charadrius collaris*
- ❏ Common Ringed Plover *Charadrius hiaticula*

## Oystercatchers – Haematopodidae
- ❏ American Oystercatcher *Haematopus palliatus*

## Stilts and Avocets – Recurvirostridae
- ❏ Black-necked Stilt *Himantopus mexicanus*
- ❏ American Avocet *Recurvirostra americana*

## Thick-knees – Burhinidae
❑ Double-striped Thick-knee *Burhinus bistriatus*

## Sandpipers and Allies – Scolopacidae
❑ Wilson's Snipe *Gallinago delicata*
❑ South American Snipe *Gallinago paraguaiae*
❑ Short-billed Dowitcher *Limnodromus griseus*
❑ Black-tailed Godwit *Limosa limosa*
❑ Hudsonian Godwit *Limosa haemastica*
❑ Marbled Godwit *Limosa fedoa*
❑ Eskimo Curlew *Numenius borealis*
❑ Whimbrel *Numenius phaeopus*
❑ Long-billed Curlew *Numenius americanus*
❑ Upland Sandpiper *Bartramia longicauda*
❑ Terek Sandpiper *Xenus cinereus*
❑ Spotted Sandpiper *Actitis macularius*
❑ Greater Yellowlegs *Tringa melanoleuca*
❑ Lesser Yellowlegs *Tringa flavipes*
❑ Spotted Redshank *Tringa erythropus*
❑ Wood Sandpiper *Tringa glareola*
❑ Common Greenshank *Tringa nebularia*
❑ Solitary Sandpiper *Tringa solitaria*
❑ Willet *Catoptrophorus semipalmatus*
❑ Ruddy Turnstone *Arenaria interpres*
❑ Red Knot *Calidris canutus*
❑ Sanderling *Calidris alba*
❑ Semipalmated Sandpiper *Calidris pusilla*
❑ Western Sandpiper *Calidris mauri*
❑ Least Sandpiper *Calidris minutilla*
❑ White-rumped Sandpiper *Calidris fuscicollis*
❑ Baird's Sandpiper *Calidris bairdii*
❑ Pectoral Sandpiper *Calidris melanotos*
❑ Curlew Sandpiper *Calidris ferruginea*
❑ Stilt Sandpiper *Calidris himantopus*
❑ Buff-breasted Sandpiper *Tryngites subruficollis*
❑ Ruff *Philomachus pugnax*
❑ Wilson's Phalarope *Phalaropus tricolor*

## Jacanas – Jacanidae
❑ Wattled Jacana *Jacana jacana*

## Skuas and Jaegers – Stercorariidae
❑ South Polar Skua *Stercorarius maccormicki*
❑ Pomarine Jaeger *Stercorarius pomarinus*
❑ Parasitic Jaeger *Stercorarius parasiticus*

## Gulls and Terns – Laridae
❑ Ring-billed Gull *Larus delawarensis*
❑ Kelp Gull *Larus dominicanus*
❑ American Herring Gull *Larus smithsonianus*
❑ Lesser Black-backed Gull *Larus fuscus*
❑ Laughing Gull *Larus atricilla*
❑ Franklin's Gull *Larus pipixcan*
❑ Black-headed Gull *Larus ridibundus*
❑ Black-legged Kittiwake *Rissa tridactyla*

❑ Sabine's Gull *Xema sabini*
❑ Brown Noddy *Anous stolidus*
❑ White Tern *Gygis alba*
❑ Sooty Tern *Onychoprion fuscata*
❑ Bridled Tern *Onychoprion anaethetus*
❑ Least Tern *Sternula antillarum*
❑ Yellow-billed Tern *Sternula superciliaris*
❑ Large-billed Tern *Phaetusa simplex*
❑ Gull-billed Tern *Gelochelidon nilotica*
❑ Caspian Tern *Hydroprogne caspia*
❑ Black Tern *Chlidonias niger*
❑ Common Tern *Sterna hirundo*
❑ Roseate Tern *Sterna dougallii*
❑ Sandwich Tern *Thalasseus sandvicensis*
❑ Royal Tern *Thalasseus maximus*

## Skimmers – Rynchopidae
❑ Black Skimmer *Rynchops niger*

## Pigeons and Doves – Columbidae
❑ Rock Pigeon *Columba livia*
❑ Scaled Pigeon *Patagioenas speciosa*
❑ Scaly-naped Pigeon *Patagioenas squamosa*
❑ Band-tailed Pigeon *Patagioenas fasciata*
❑ Pale-vented Pigeon *Patagioenas cayennensis*
❑ Common Ground Dove *Columbina passerina*
❑ Plain-breasted Ground Dove *Columbina minuta*
❑ Ruddy Ground Dove *Columbina talpacoti*
❑ Scaled Dove *Columbina squammata*
❑ Blue Ground Dove *Claravis pretiosa*
❑ Eared Dove *Zenaida auriculata*
❑ White-tipped Dove *Leptotila verreauxi*
❑ Grey-fronted Dove *Leptotila rufaxilla*
❑ Lined Quail-Dove *Geotrygon linearis*
❑ Ruddy Quail-Dove *Geotrygon montana*

## Macaws and Parrots – Psittacidae
❑ Blue-and-yellow Macaw *Ara ararauna*
❑ Scarlet Macaw *Ara macao*
❑ Red-bellied Macaw *Orthopsittaca manilata*
❑ Brown-throated Parakeet *Aratinga pertinax*
❑ Green-rumped Parrotlet *Forpus passerinus*
❑ Lilac-tailed Parrotlet *Touit batavicus*
❑ Scarlet-shouldered Parrotlet *Touit huetii*
❑ Blue-headed Parrot *Pionus menstruus*
❑ Yellow-crowned Amazon *Amazona ochrocephala*
❑ Orange-winged Amazon *Amazona amazonica*

## Cuckoos – Cuculidae
❑ Black-billed Cuckoo *Coccyzus erythropthalmus*
❑ Yellow-billed Cuckoo *Coccyzus americanus*
❑ Mangrove Cuckoo *Coccyzus minor*
❑ Dark-billed Cuckoo *Coccyzus melacoryphus*
❑ Squirrel Cuckoo *Piaya cayana*
❑ Little Cuckoo *Piaya minuta*

- ❑ Striped Cuckoo *Tapera naevia*
- ❑ Greater Ani *Crotophaga major*
- ❑ Smooth-billed Ani *Crotophaga ani*

## Barn Owls – Tytonidae
- ❑ Barn Owl *Tyto alba*

## Owls – Strigidae
- ❑ Tropical Screech Owl *Megascops choliba*
- ❑ Spectacled Owl *Pulsatrix perspicillata*
- ❑ Mottled Owl *Ciccaba virgata*
- ❑ Ferruginous Pygmy Owl *Glaucidium brasilianum*
- ❑ Burrowing Owl *Athene cunicularia*
- ❑ Striped Owl *Asio clamator*
- ❑ Short-eared Owl *Asio flammeus*

## Oilbird – Steatornithidae
- ❑ Oilbird *Steatornis caripensis*

## Potoos – Nyctibiidae
- ❑ Common Potoo *Nyctibius griseus*

## Nightjars and Nighthawks – Caprimulgidae
- ❑ Short-tailed Nighthawk *Lurocalis semitorquatus*
- ❑ Lesser Nighthawk *Chordeiles acutipennis*
- ❑ Nacunda Nighthawk *Podager nacunda*
- ❑ Common Pauraque *Nyctidromus albicollis*
- ❑ Rufous Nightjar *Caprimulgus rufus*
- ❑ White-tailed Nightjar *Caprimulgus cayennensis*

## Swifts – Apodidae
- ❑ Chestnut-collared Swift *Streptoprocne rutila*
- ❑ White-collared Swift *Streptoprocne zonaris*
- ❑ Band-rumped Swift *Chaetura spinicaudus*
- ❑ Grey-rumped Swift *Chaetura cinereiventris*
- ❑ Chapman's Swift *Chaetura chapmani*
- ❑ Short-tailed Swift *Chaetura brachyura*
- ❑ Fork-tailed Palm Swift *Tachornis squamata*
- ❑ Lesser Swallow-tailed Swift *Panyptila cayennensis*

## Hummingbirds – Trochilidae
- ❑ Rufous-breasted Hermit *Glaucis hirsutus*
- ❑ Little Hermit *Phaethornis longuemareus*
- ❑ Green Hermit *Phaethornis guy*
- ❑ White-tailed Sabrewing *Campylopterus ensipennis*
- ❑ White-necked Jacobin *Florisuga mellivora*
- ❑ Brown Violetear *Colibri delphinae*
- ❑ Green-throated Mango *Anthracothorax viridigula*
- ❑ Black-throated Mango *Anthracothorax nigricollis*
- ❑ Ruby Topaz *Chrysolampis mosquitus*
- ❑ Tufted Coquette *Lophornis ornatus*
- ❑ Blue-chinned Sapphire *Chlorestes notata*
- ❑ Blue-tailed Emerald *Chlorostilbon mellisugus*
- ❑ White-tailed Goldenthroat *Polytmus guainumbi*

- ❑ White-chested Emerald *Amazilia chionopectus*
- ❑ Copper-rumped Hummingbird *Amazilia tobaci*
- ❑ Long-billed Starthroat *Heliomaster longirostris*
- ❑ Rufous-shafted Woodstar *Chaetocercus jourdanii*

## Trogons – Trogonidae
- ❑ Amazonian White-tailed Trogon *Trogon viridis*
- ❑ Amazonian Violaceous Trogon *Trogon violaceus*
- ❑ Collared Trogon *Trogon collaris*

## Kingfishers – Alcedinidae
- ❑ Ringed Kingfisher *Megaceryle torquata*
- ❑ Belted Kingfisher *Megaceryle alcyon*
- ❑ Amazon Kingfisher *Chloroceryle amazona*
- ❑ Green Kingfisher *Chloroceryle americana*
- ❑ Pygmy Kingfisher *Chloroceryle aenea*

## Motmots – Momotidae
- ❑ Blue-crowned Motmot *Momotus momota*

## Jacamars – Galbulidae
- ❑ Rufous-tailed Jacamar *Galbula ruficauda*

## Toucans – Ramphastidae
- ❑ Channel-billed Toucan *Ramphastos vitellinus*

## Woodpeckers – Picidae
- ❑ Red-crowned Woodpecker *Melanerpes rubricapillus*
- ❑ Red-rumped Woodpecker *Veniliornis kirkii*
- ❑ Golden-olive Woodpecker *Piculus rubiginosus*
- ❑ Chestnut Woodpecker *Celeus elegans*
- ❑ Lineated Woodpecker *Dryocopus lineatus*
- ❑ Crimson-crested Woodpecker *Campephilus melanoleucos*

## Ovenbirds – Furnariinae
- ❑ Pale-breasted Spinetail *Synallaxis albescens*
- ❑ Stripe-breasted Spinetail *Synallaxis cinnamomea*
- ❑ Yellow-chinned Spinetail *Certhiaxis cinnamomeus*
- ❑ Grey-throated Leaftosser *Sclerurus albigularis*
- ❑ Streaked Xenops *Xenops rutilans*

## Woodcreepers – Dendrocolaptinae
- ❑ Plain-brown Woodcreeper *Dendrocincla fuliginosa*
- ❑ Olivaceous Woodcreeper *Sittasomus griseicapillus*
- ❑ Straight-billed Woodcreeper *Xiphorhynchus picus*
- ❑ Cocoa Woodcreeper *Xiphorhynchus susurrans*
- ❑ Streak-headed Woodcreeper *Lepidocolaptes souleyetii*

## Antbirds – Thamnophilidae
- ❑ Great Antshrike *Taraba major*
- ❑ Black-crested Antshrike *Sakesphorus canadensis*
- ❑ Barred Antshrike *Thamnophilus doliatus*
- ❑ Plain Antvireo *Dysithamnus mentalis*

- ❏ White-flanked Antwren *Myrmotherula axillaris*
- ❏ White-fringed Antwren *Formicivora grisea*
- ❏ Silvered Antbird *Sclateria naevia*
- ❏ White-bellied Antbird *Myrmeciza longipes*

## Antthrushes and Antpittas – Formicariidae
- ❏ Black-faced Antthrush *Formicarius analis*
- ❏ Scaled Antpitta *Grallaria guatimalensis*

## Tyrant Flycatchers – Tyrannidae
- ❏ Forest Elaenia *Myiopagis gaimardii*
- ❏ Yellow-bellied Elaenia *Elaenia flavogaster*
- ❏ Small-billed Elaenia *Elaenia parvirostris*
- ❏ Slaty Elaenia *Elaenia strepera*
- ❏ Lesser Elaenia *Elaenia chiriquensis*
- ❏ Southern Beardless Tyrannulet *Camptostoma obsoletum*
- ❏ Mouse-coloured Tyrannulet *Phaeomyias murina*
- ❏ Crested Doradito *Pseudocolopteryx sclateri*
- ❏ Olive-striped Flycatcher *Mionectes olivaceus*
- ❏ Ochre-bellied Flycatcher *Mionectes oleagineus*
- ❏ Slaty-capped Flycatcher *Leptopogon superciliaris*
- ❏ Northern Scrub Flycatcher *Sublegatus arenarum*
- ❏ Short-tailed Pygmy Tyrant *Myiornis ecaudatus*
- ❏ Spotted Tody-Flycatcher *Todirostrum maculatum*
- ❏ Yellow-olive Flycatcher *Tolmomyias sulphurescens*
- ❏ Yellow-breasted Flycatcher *Tolmomyias flaviventris*
- ❏ White-throated Spadebill *Platyrinchus mystaceus*
- ❏ Bran-coloured Flycatcher *Myiophobus fasciatus*
- ❏ Euler's Flycatcher *Lathrotriccus euleri*
- ❏ Fuscous Flycatcher *Cnemotriccus fuscatus*
- ❏ Olive-sided Flycatcher *Contopus cooperi*
- ❏ Tropical Pewee *Contopus cinereus*
- ❏ Pied Water Tyrant *Fluvicola pica*
- ❏ White-headed Marsh Tyrant *Arundinicola leucocephala*
- ❏ Piratic Flycatcher *Legatus leucophaius*
- ❏ Great Kiskadee *Pitangus sulphuratus*
- ❏ Streaked Flycatcher *Myiodynastes maculatus*
- ❏ Boat-billed Flycatcher *Megarynchus pitangua*
- ❏ Sulphury Flycatcher *Tyrannopsis sulphurea*
- ❏ Variegated Flycatcher *Empidonomus varius*
- ❏ Tropical Kingbird *Tyrannus melancholicus*
- ❏ Fork-tailed Flycatcher *Tyrannus savana*
- ❏ Grey Kingbird *Tyrannus dominicensis*
- ❏ Dusky-capped Flycatcher *Myiarchus tuberculifer*
- ❏ Swainson's Flycatcher *Myiarchus swainsoni*
- ❏ Venezuelan Flycatcher *Myiarchus venezuelensis*
- ❏ Brown-crested Flycatcher *Myiarchus tyrannulus*
- ❏ Bright-rumped Attila *Attila spadiceus*

## Cotingas – Cotingidae
- ❏ White Bellbird *Procnias albus*
- ❏ Bearded Bellbird *Procnias averano*

## Manakins – Pipridae
- ❏ White-bearded Manakin *Manacus manacus*
- ❏ Blue-backed Manakin *Chiroxiphia pareola*
- ❏ Golden-headed Manakin *Pipra erythrocephala*

## Tityras and Becards – Tityridae
- ❏ Black-tailed Tityra *Tityra cayana*
- ❏ White-winged Becard *Pachyramphus polychopterus*

## Vireos – Vireonidae
- ❏ Rufous-browed Peppershrike *Cyclarhis gujanensis*
- ❏ Yellow-throated Vireo *Vireo flavifrons*
- ❏ White-eyed Vireo *Vireo griseus*
- ❏ Red-eyed Vireo *Vireo olivaceus*
- ❏ Black-whiskered Vireo *Vireo altiloquus*
- ❏ Golden-fronted Greenlet *Hylophilus aurantiifrons*
- ❏ Scrub Greenlet *Hylophilus flavipes*

## Swallows and Martins – Hirundinidae
- ❏ White-winged Swallow *Tachycineta albiventer*
- ❏ Caribbean Martin *Progne dominicensis*
- ❏ Grey-breasted Martin *Progne chalybea*
- ❏ Blue-and-white Swallow *Notiochelidon cyanoleuca*
- ❏ Southern Rough-winged Swallow *Stelgidopteryx ruficollis*
- ❏ Bank Swallow *Riparia riparia*
- ❏ Barn Swallow *Hirundo rustica*
- ❏ Cliff Swallow *Petrochelidon pyrrhonota*

## Wrens – Troglodytidae
- ❏ Southern House Wren *Troglodytes musculus*
- ❏ Rufous-breasted Wren *Thryothorus rutilus*

## Gnatwrens – Sylviidae
- ❏ Long-billed Gnatwren *Ramphocaenus melanurus*

## Thrushes – Turdidae
- ❏ Orange-billed Nightingale-Thrush *Catharus aurantiirostris*
- ❏ Veery *Catharus fuscescens*
- ❏ Grey-cheeked Thrush *Catharus minimus*
- ❏ Yellow-legged Thrush *Turdus flavipes*
- ❏ Cocoa Thrush *Turdus fumigatus*
- ❏ Bare-eyed Thrush *Turdus nudigenis*
- ❏ White-necked Thrush *Turdus albicollis*

## Mockingbirds – Mimidae
- ❏ Tropical Mockingbird *Mimus gilvus*

## Wagtails – Motacillidae
- ❏ White Wagtail *Motacilla alba*

## Tanagers – Thraupidae
- ❏ White-shouldered Tanager *Tachyphonus luctuosus*
- ❏ White-lined Tanager *Tachyphonus rufus*

- ❑ Silver-beaked Tanager *Ramphocelus carbo*
- ❑ Blue-grey Tanager *Thraupis episcopus*
- ❑ Palm Tanager *Thraupis palmarum*
- ❑ Blue-capped Tanager *Thraupis cyanocephala*
- ❑ Turquoise Tanager *Tangara mexicana*
- ❑ Speckled Tanager *Tangara guttata*
- ❑ Bay-headed Tanager *Tangara gyrola*
- ❑ Swallow Tanager *Tersina viridis*
- ❑ Blue Dacnis *Dacnis cayana*
- ❑ Purple Honeycreeper *Cyanerpes caeruleus*
- ❑ Red-legged Honeycreeper *Cyanerpes cyaneus*
- ❑ Green Honeycreeper *Chlorophanes spiza*
- ❑ Bicoloured Conebill *Conirostrum bicolor*

## Genera Incertae Sedis
- ❑ Bananaquit *Coereba flaveola*
- ❑ Sooty Grassquit *Tiaris fuliginosus*
- ❑ Black-faced Grassquit *Tiaris bicolor*

## Seedeaters and Allies – Emberizidae
- ❑ Orange-fronted Yellow Finch *Sicalis columbiana*
- ❑ Saffron Finch *Sicalis flaveola*
- ❑ Grassland Yellow Finch *Sicalis luteola*
- ❑ Blue-black Grassquit *Volatinia jacarina*
- ❑ Slate-coloured Seedeater *Sporophila schistacea*
- ❑ Grey Seedeater *Sporophila intermedia*
- ❑ Wing-barred Seedeater *Sporophila americana*
- ❑ Lesson's Seedeater *Sporophila bouvronides*
- ❑ Yellow-bellied Seedeater *Sporophila nigricollis*
- ❑ Ruddy-breasted Seedeater *Sporophila minuta*
- ❑ Chestnut-bellied Seed Finch *Oryzoborus angolensis*
- ❑ Large-billed Seed Finch *Oryzoborus crassirostris*
- ❑ Red-capped Cardinal *Paroaria gularis*

## Cardinals, Grosbeaks, Saltators and Allies – Cardinalidae
- ❑ Rose-breasted Grosbeak *Pheucticus ludovicianus*
- ❑ Greyish Saltator *Saltator coerulescens*
- ❑ Streaked Saltator *Saltator striatipectus*
- ❑ Indigo Bunting *Passerina cyanea*
- ❑ Dickcissel *Spiza americana*

## Genera Incertae Sedis
- ❑ Highland Hepatic Tanager *Piranga flava*
- ❑ Summer Tanager *Piranga rubra*
- ❑ Scarlet Tanager *Piranga olivacea*
- ❑ Crowned Ant Tanager *Habia rubica*

## New World Warblers – Parulidae
- ❑ Golden-winged Warbler *Vermivora chrysoptera*
- ❑ Northern Parula *Parula americana*
- ❑ Tropical Parula *Parula pitiayumi*
- ❑ Chestnut-sided Warbler *Dendroica pensylvanica*
- ❑ Yellow Warbler *Dendroica petechia*

- ❑ Blackpoll Warbler *Dendroica striata*
- ❑ Bay-breasted Warbler *Dendroica castanea*
- ❑ Blackburnian Warbler *Dendroica fusca*
- ❑ Magnolia Warbler *Dendroica magnolia*
- ❑ Cerulean Warbler *Dendroica cerulea*
- ❑ Cape May Warbler *Dendroica tigrina*
- ❑ Black-throated Blue Warbler *Dendroica caerulescens*
- ❑ Yellow-rumped Warbler *Dendroica coronata*
- ❑ Black-throated Green Warbler *Dendroica virens*
- ❑ Prairie Warbler *Dendroica discolor*
- ❑ American Redstart *Setophaga ruticilla*
- ❑ Black-and-white Warbler *Mniotilta varia*
- ❑ Prothonotary Warbler *Protonotaria citrea*
- ❑ Ovenbird *Seiurus aurocapilla*
- ❑ Northern Waterthrush *Seiurus noveboracensis*
- ❑ Common Yellowthroat *Geothlypis trichas*
- ❑ Masked Yellowthroat *Geothlypis aequinoctialis*
- ❑ Hooded Warbler *Wilsonia citrina*
- ❑ Canada Warbler *Wilsonia canadensis*
- ❑ Golden-crowned Warbler *Basileuterus culicivorus*

## New World Blackbirds – Icteridae
- ❑ Crested Oropendola *Psarocolius decumanus*
- ❑ Yellow-rumped Cacique *Cacicus cela*
- ❑ Moriche Oriole *Icterus chrysocephalus*
- ❑ Orchard Oriole *Icterus spurius*
- ❑ Baltimore Oriole *Icterus galbula*
- ❑ Yellow Oriole *Icterus nigrogularis*
- ❑ Yellow-hooded Blackbird *Chrysomus icterocephalus*
- ❑ Giant Cowbird *Molothrus oryzivorus*
- ❑ Shiny Cowbird *Molothrus bonariensis*
- ❑ Carib Grackle *Quiscalus lugubris*
- ❑ Red-breasted Blackbird *Sturnella militaris*
- ❑ Red-winged Blackbird *Agelaius phoeniceus*
- ❑ Bobolink *Dolichonyx oryzivorus*

## Siskins – Carduelinae
- ❑ Red Siskin *Carduelis cucullata*
- ❑ Lesser Goldfinch *Carduelis psaltria*

## Euphonias – Euphoniinae
- ❑ Trinidad Euphonia *Euphonia trinitatis*
- ❑ Violaceous Euphonia *Euphonia violacea*
- ❑ Golden-rumped Euphonia *Euphonia cyanocephala*

## Waxbills and Allies – Estrildidae
- ❑ Common Waxbill *Estrilda astrild*

# TRINIDAD & TOBAGO RARE BIRDS COMMITTEE

Formed in 1995, the primary function of the Committee is to assess submitted descriptions and/or photographs of rare species reported from the islands. The Review List details those species for which documentary evidence is sought. In view of the increased birdwatcher coverage of Trinidad & Tobago, this list is regularly monitored and will periodically be amended to reflect any change in distribution and occurrence.

Current, at the time of writing, members of the Committee are Martyn Kenefick (Hon. Sec.), Geoffrey Gomes, Courtenay Rooks and Graham White, all based on Trinidad, Floyd Hayes and Bill Murphy, in the USA, and Richard ffrench in the UK.

## Review List

The Trinidad & Tobago Rare Bird Committee wishes to review submitted reports of the species listed below:

**Screamers:** Horned Screamer *Anhima cornuta*. **Ducks and Geese:** Fulvous Whistling Duck *Dendrocygna bicolor* (Tobago only), Snow Goose *Chen caerulescens*, Comb Duck *Sarkidiornis melanotos*, American Wigeon *Anas americana*, Northern Shoveler *A. clypeata*, Northern Pintail *A. acuta*, Green-winged Teal *A. carolinensis*, Southern Pochard *Netta erythrophthalma*, Ring-necked Duck *Aythya collaris*, Lesser Scaup *A. affinis*, Masked Duck *Nomonyx dominicus*. **Chachalacas and Guans:** Trinidad Piping Guan *Pipile pipile* (away from Grande Rivière). **Petrels and Shearwaters:** Bulwer's Petrel *Bulweria bulwerii*, Cory's Shearwater *Calonectris diomedea*, Great Shearwater *Puffinus gravis*, Sooty Shearwater *P. griseus*, Manx Shearwater *P. puffinus*, Audubon's Shearwater *P. lherminieri* (Trinidad only). **Storm-Petrels:** Wilson's Storm-Petrel *Oceanites oceanicus*, **Tropicbirds:** White-tailed Tropicbird *Phaethon lepturus*, Red-billed Tropicbird *P. aethereus* (Trinidad only). **Boobies:** Masked Booby *Sula dactylatra* (away from St Giles Rocks and Little Tobago), Red-footed Booby *S. sula* (Trinidad only), Northern Gannet *Morus bassanus*. **Herons:** Rufescent Tiger Heron *Tigrisoma lineatum*, Agami Heron *Agamia agami*, Grey Heron *Ardea cinerea*, Purple Heron *A. purpurea*, Cocoi Heron *A. cocoi* (Tobago only), Western Reef Heron *Egretta gularis*, Reddish Egret *E. rufescens*, Green Heron *Butorides virescens* (Trinidad only), Striated Heron *B. striata* (Tobago only). **Ibises and Spoonbills:** White Ibis *Eudocimus albus*, Scarlet Ibis *E. ruber* (Tobago only), Glossy Ibis *Plegadis falcinellus*, Eurasian Spoonbill *Platalea leucorodia*, Roseate Spoonbill *P. ajaja*. **Storks:** Maguari Stork *Ciconia maguari*, Jabiru *Jabiru mycteria*, Wood Stork *Mycteria americana*. **New World Vultures:** King Vulture *Sarcoramphus papa*. **Flamingos:** Greater Flamingo *Phoenicopterus ruber*. **Hawks, Kites and Eagles:** Hook-billed Kite *Chondrohierax uncinatus*, Swallow-tailed Kite *Elanoides forficatus* (Tobago only), White-tailed Kite *Elanus leucurus*, Snail Kite *Rostrhamus sociabilis*, Crane Hawk *Geranospiza caerulescens*, Grey-lined Hawk *Asturina nitida* (Tobago only), Common Black Hawk *Buteogallus anthracinus* (Tobago only), Rufous Crab Hawk *B. aequinoctialis*, Great Black Hawk *B. urubitinga* (Trinidad only), Savanna Hawk *B. meridionalis* (Tobago only), Black-collared Hawk *Busarellus nigricollis*, Short-tailed Hawk *Buteo brachyurus* (Tobago only), Swainson's Hawk *B. swainsoni*, White-tailed Hawk *B. albicaudatus*, Black Hawk-Eagle *Spizaetus tyrannus*, Ornate Hawk-Eagle *S. ornatus* (Tobago only). **Caracaras and Falcons:** Northern Crested Caracara *Caracara cheriway*, Common Kestrel *Falco tinnunculus*, American Kestrel *F. sparverius*, Bat Falcon *F. rufigularis* (Tobago only), Orange-breasted Falcon *F. deiroleucus*, Aplomado Falcon *F. femoralis*. **Rails, Crakes and Gallinules:** Rufous-necked Wood Rail *Aramides axillaries*, Yellow-breasted Crake *Porzana flaviventer* (Tobago only), Ash-throated Crake *P. albicollis*, Paint-billed Crake *Neocrex erythrops*, Spotted Rail *Pardirallus maculates*, Azure Gallinule *Porphyrio flavirostris* (away from Nariva swamp), Caribbean Coot *Fulica caribaea*, American Coot *F. americana*. **Sungrebe:** Sungrebe *Heliornis fulica*. **Lapwings and Plovers:** Pied Lapwing *Vanellus cayanus*, Wilson's Plover *Charadrius wilsonia* (Tobago only), Killdeer *C. vociferous*, Snowy Plover *C. alexandrinus*, Common Ringed Plover *C. hiaticula*. **Oystercatchers:** American Oystercatcher *Haematopus palliatus*. **Stilts and Avocets:** Black-necked Stilt *Himantopus mexicanus* (Tobago only), American Avocet *Recurvirostra americana*. **Thick-knees:** Double-striped Thick-knee *Burhinus bistriatus*. **Sandpipers:** Black-tailed Godwit *Limosa limosa*, Hudsonian Godwit *L. haemastica* (Tobago only), Marbled Godwit *L. fedoa*, Eskimo Curlew *Numenius borealis*, Long-billed Curlew *N. americanus*, Upland Sandpiper *Bartramia longicauda*, Terek Sandpiper *Xenus cinereus*, Spotted Redshank *Tringa erythropus*, Wood Sandpiper *T. glareola*, Common Greenshank *T. nebularia*, Baird's Sandpiper *Calidris bairdii*, Curlew Sandpiper *C. ferruginea*, Buff-breasted Sandpiper *Tryngites subruficollis*, Ruff *Philomachus pugnax*, Wilson's Phalarope *Phalaropus tricolor*. **Skuas:** Great Skua *Stercorarius skua*, South Polar Skua *S. maccormicki*, Pomarine Jaeger *S. pomarinus*. **Gulls and Terns:** Ring-billed Gull *Larus delawarensis*, Kelp Gull *L. dominicanus*, American Herring Gull *L. smithsonianus*, Lesser Black-backed

Gull *L. fuscus* (Tobago only), Franklin's Gull *L. pipixcan*, Black-headed Gull *L. ridibundus*, Black-legged Kittiwake *Rissa tridactyla*, Sabine's Gull *Xema sabini*, White Tern *Gygis alba*, Caspian Tern *Hydroprogne caspia*. **Skimmers:** Black Skimmer *Rynchops niger* (Tobago only). **Pigeons and Doves:** Scaled Dove *Columbina squammata*, Blue Ground Dove *Claravis pretiosa*, Scaly-naped Pigeon *Patagioenas squamosa*, Band-tailed Pigeon *P. fasciata*, Lined Quail Dove *Geotrygon linearis*. **Macaws and Parrots:** Blue-and-yellow Macaw *Ara ararauna*, Scarlet Macaw *A. macao*, Brown-throated Parakeet *Aratinga pertinax*, Scarlet-shouldered Parrotlet *Touit huetii*. **Cuckoos:** Black-billed Cuckoo *Coccyzus erythropthalmus*, Dark-billed Cuckoo *C. melacoryphus*. **Owls:** Mottled Owl *Ciccaba virgata*, Burrowing Owl *Athene cunicularia*, Striped Owl *Asio clamator*, Short-eared Owl *Asio flammeus*. **Oilbird:** Oilbird *Steatornis caripensis* (Tobago only). **Nightjars and Nighthawks:** Lesser Nighthawk *Chordeiles acutipennis* (Tobago only), Nacunda Nighthawk *Podager nacunda* (Tobago only). **Swifts:** White-collared Swift *Streptoprocne zonaris* (Tobago only), Chapman's Swift *Chaetura chapmani*, Fork-tailed Palm Swift *Tachornis squamata* (Tobago only). **Hummingbirds:** Brown Violetear *Colibri delphinae* (Tobago only), Blue-chinned Sapphire *Chlorestes notata* (Tobago only), Rufous-shafted Woodstar *Chaetocercus jourdanii*. **Kingfishers:** Amazon Kingfisher *Chloroceryle amazona*. **Antthrushes and Antpittas:** Scaled Antpitta *Grallaria guatimalensis*. **Tyrant Flycatchers:** Small-billed Elaenia *Elaenia parvirostris*, Slaty Elaenia *E. strepera*, Lesser Elaenia *E. chiriquensis*, Crested Doradito *Pseudocolopteryx sclateri*, Olive-striped Flycatcher *Mionectes olivaceus*, Spotted Tody-flycatcher *Todirostrum maculatum* (away from Oropouche and Icacos), Variegated Flycatcher *Empidonomus varius*, Swainson's Flycatcher *Myiarchus swainsoni*. **Cotingas:** White Bellbird *Procnias albus*. **Vireos:** Yellow-throated Vireo *Vireo flavifrons*, White-eyed Vireo *V. griseus*, Black-whiskered Vireo *V. altiloquus*. **Swallows and Martins:** Caribbean Martin *Progne dominicensis* (Trinidad only), Grey-breasted Martin *P. chalybea* (Tobago only), Southern Rough-winged Swallow *Stelgidopteryx ruficollis* (Tobago only), Bank Swallow *Riparia riparia* (Tobago only), Cliff Swallow *Petrochelidon pyrrhonota*. **Thrushes:** Veery *Catharus fuscescens*, Grey-cheeked Thrush *C. minimus*. **Wagtails:** White Wagtail *Motacilla alba*. **Tanagers:** Purple Honeycreeper *Cyanerpes caeruleus* (Tobago only). **Seedeaters and Allies:** Orange-fronted Yellow Finch *Sicalis columbiana*, Grassland Yellow Finch *S. luteola* (away from Aripo Agriculture Station), Slate-coloured Seedeater *Sporophila schistacea*, Grey Seedeater *S. intermedia*, Ring-necked Seedeater *S. insularis*, Wing-barred Seedeater *S. americana*, Lesson's Seedeater *S. bouvronides*, Yellow-bellied Seedeater *S. nigricollis*, Ruddy-breasted Seedeater *S. minuta* (away from Aripo Agriculture Station), Chestnut-bellied Seed Finch *Oryzoborus angolensis*, Large-billed Seed Finch *O. crassirostris*. **Cardinals, Grosbeaks and Saltators:** Rose-breasted Grosbeak *Pheucticus ludovicianus*, Streaked Saltator *Saltator striatipectus* (away from Bocas Islands and Chaguaramas Peninsula), Indigo Bunting *Passerina cyanea*, Dickcissel *Spiza americana* (Tobago only). **Genera Incertae Sedis:** Summer Tanager *Piranga rubra*, Scarlet Tanager *P. olivacea*. **New World Warblers:** Golden-winged Warbler *Vermivora chrysoptera*, Northern Parula *Parula americana*, Tropical Parula *P. pitiayumi* (Tobago only), Chestnut-sided Warbler *Dendroica pensylvanica*, Bay-breasted Warbler *D. castanea*, Blackburnian Warbler *D. fusca*, Magnolia Warbler *D. magnolia*, Cerulean Warbler *D. caerulea*, Cape May Warbler *D. tigrina*, Black-throated Blue Warbler *D. caerulescens*, Yellow-rumped Warbler *D. coronata*, Black-throated Green Warbler *D. virens*, Prairie Warbler *D. discolor*, Black-and-white Warbler *Mniotilta varia*, Ovenbird *Seiurus aurocapilla*, Common Yellowthroat *Geothlypis trichas*, Hooded Warbler *Wilsonia citrina*, Canada Warbler *W. canadensis*. **New World Blackbirds:** Moriche Oriole *Icterus chrysocephalus* (away from Aripo savannah), Orchard Oriole *I. spurious*, Baltimore Oriole *I. galbula*, Red-winged Blackbird *Agelaius phoeniceus*, Red-breasted Blackbird *Sturnella militaris* (Tobago only), Bobolink *Dolichonyx oryzivorus*. **Siskins:** Red Siskin *Carduelis cucullata*, Lesser Goldfinch *C. psaltria*. **Euphonias:** Golden-rumped Euphonia *Euphonia cyanocephala*.

Submissions can be made to Martyn Kenefick, 36 Newalloville Avenue, San Juan, Trinidad, or by e-mail: martynkenefick@hotmail.com

# INDEX

All entries refer to Plate numbers, not pages.

carolina, *Porzana* 31
carolinensis, *Anas* 4
caspia, *Hydroprogne* 47
castanea, *Dendroica* 100
*CATHARTES* 17
*CATHARUS* 88
*CATOPTROPHORUS* 39
cayana, *Dacnis* 91
cayana, *Piaya* 58
cayana, *Tityra* 83
cayanensis, *Leptodon* 19
cayanus, *Vanellus* 34
cayennensis, *Caprimulgus* 61
cayennensis, *Panyptila* 63
cayennensis, *Patagioenas* 52
cela, *Cacicus* 104
*CELEUS* 71
*CERTHIAXIS* 72
cerulea, *Dendroica* 102
Chachalaca, Rufous-vented 1
*CHAETOCERCUS* 66
*CHAETURA* 63
chalybea, *Progne* 86
chapmani, *Chaetura* 63
*CHARADRIUS* 35
*CHEN* 2
chilensis, *Vanellus* 34
chimachima, *Milvago* 27
chionopectus, *Amazilia* 66
chiriquensis, *Elaenia* 77
*CHIROXIPHIA* 84
*CHLIDONIAS* 49
*CHLORESTES* 65
*CHLOROCERYLE* 70
*CHLOROPHANES* 92
chloropus, *Gallinula* 32
*CHLOROSTILBON* 65
choliba, *Megascops* 59
*CHONDROHIERAX* 18
*CHORDEILES* 62
chrysocephalus, *Icterus* 104
*CHRYSOLAMPIS* 64
*CHRYSOMUS* 104
chrysoptera, *Vermivora* 99
*CICCABA* 60
*CICONIA* 16
cinerea, *Ardea* 13
cinereiventris, *Chaetura* 63
cinereus, *Contopus* 79
cinereus, *Xenus* 39
cinnamomea, *Synallaxis* 72
cinnamomeus, *Certhiaxis* 72
*CIRCUS* 21
citrea, *Protonotaria* 102
citrina, *Wilsonia* 103
clamator, *Asio* 60
*CLARAVIS* 53
clypeata, *Anas* 3
*CNEMOTRICCUS* 79
*COCCYZUS* 57

*COCHLEARIUS* 11
cochlearius, *Cochlearius* 11
cocoi, *Ardea* 13
*COEREBA* 92
coerulescens, *Saltator* 97
*COLIBRI* 64
collaris, *Aythya* 5
collaris, *Charadrius* 35
collaris, *Trogon* 67
*COLUMBA* 52
columbarius, *Falco* 29
columbiana, *Sicalis* 93
*COLUMBINA* 53
Conebill, Bicoloured 92
*CONIROSTRUM* 92
*CONTOPUS* 79
cooperi, *Contopus* 79
Coot, American 32
    Caribbean 32
Coquette, Tufted 66
*CORAGYPS* 17
Cormorant, Neotropic 6
cornuta, *Anhima* 1
coronata, *Dendroica* 101
Cowbird, Giant 105
    Shiny 105
Crake, Ash-throated 31
    Grey-breasted 31
    Paint-billed 31
    Yellow-breasted 31
crassirostris, *Oryzoborus* 96
*CROTOPHAGA* 58
*CRYPTURELLUS* 1
Cuckoo, Black-billed 57
    Dark-billed 57
    Little 58
    Mangrove 57
    Squirrel 58
    Striped 57
    Yellow-billed 57
cucullata, *Carduelis* 107
culicivorus, *Basileuterus* 102
cunicularia, *Athene* 60
Curlew, Eskimo 37
    Long-billed 37
cyanea, *Passerina* 96
*CYANERPES* 92
cyaneus, *Cyanerpes* 92
cyanocephala, *Euphonia* 107
cyanocephala, *Thraupis* 91
cyanoleuca, *Notiochelidon* 86
*CYCLARHIS* 85
*DACNIS* 91
Dacnis, Blue 91
decaocto, *Streptopelia* 54
decumanus, *Psarocolius* 104
deiroleucus, *Falco* 28
delawarensis, *Larus* 44
delicata, *Gallinago* 36
delphinae, *Colibri* 64

*DENDROCINCLA* 73
*DENDROCYGNA* 2
*DENDROICA* 99–102
Dickcissel 96
diomedea, *Calonectris* 7
discolor, *Dendroica* 101
discors, *Anas* 3
doliatus, *Thamnophilus* 74
*DOLICHONYX* 106
dominica, *Pluvialis* 34
dominicanus, *Larus* 44
dominicensis, *Progne* 86
dominicensis, *Tyrannus* 81
dominicus, *Nomonyx* 4
dominicus, *Tachybaptus* 6
Doradito, Crested 76
dougallii, *Sterna* 48
Dove, Blue Ground 53
    Common Ground 53
    Eared 52
    Eurasian Collared 54
    Grey-fronted 54
    Plain-breasted Ground 53
    Ruddy Ground 53
    Scaled 53
    White-tipped 54
Dowitcher, Short-billed 36
*DRYOCOPUS* 71
Duck, Black-bellied Whistling 2
    Comb 2
    Fulvous Whistling 2
    Masked 4
    Muscovy 4
    Ring-necked 5
    White-faced Whistling 2
*DYSITHAMNUS* 75
ecaudatus, *Myiornis* 76
Egret, Cattle 12
    Great 13
    Little 12
    Reddish 14
    Snowy 12
*EGRETTA* 12, 14
*ELAENIA* 77
Elaenia, Forest 77
    Lesser 77
    Slaty 77
    Small-billed 77
    Yellow-bellied 77
*ELANOIDES* 19
*ELANUS* 19
elegans, *Celeus* 71
Emerald, Blue-tailed 65
    White-chested 66
*EMPIDONOMUS* 80
ensipennis, *Campylopterus* 64
episcopus, *Thraupis* 90
erythrocephala, *Pipra* 84
erythrophthalma, *Netta* 5
erythrophthalmus, *Coccyzus* 57